HORRIE THE WOG-DOG

-

*With the A.I.F. in
Egypt, Greece, Crete
And Palestine*

Written from the Diary of
J.B. MOODY, PRIVATE VX13091, A.I.F.

ION IDRIESS

ETT IMPRINT

Exile Bay

This 11th edition published as an Imprint Classic, Exile Bay 2024

ETT IMPRINT
PO Box R1906
Royal Exchange NSW 1225 Australia

First published by Angus & Robertson Publishers, 1945.
Reprinted 1948 (three), 1949, 1951, 1955.

First published by Bobbs Merrill in the USA as *Dog of the Desert*, 1945.
Published by ETT Imprint in 2017. Reprinted 2018, 2020.

© Idriess Enterprises Pty Ltd, 1945, 2017

This book is copyright. Apart from any fair dealing for the purposes of private study, research, criticism or review, as permitted under the Copyright Act, no part may be reproduced by any process without written permission. Inquiries should be addressed to the publishers.

ISBN 978-1-925416-98-5 (ebook)
ISBN 978-1-923024-92-2 (paper)

Design by Hanna Gotlieb
Cover design by Tom Thompson

CONTENTS

1.	INTRODUCING THE DESERT, THE PUP AND THE REBELS	3
2.	THE WOG-DOG ENLISTS	8
3.	THE REBELS	13
4.	MURCHIE'S ASP	21
5.	ONE NIGHT IN ALEXANDRIA	25
6.	DEEP LAID PLANS	30
7.	THE WOG-DOG EMBARKS FOR THE FRONT	39
8.	THE WOG-DOG UNDER FIRE	47
9.	THE WOG-DOG CONQUERS GREECE	57
10.	THE WOG-DOG SAVES LIVES	66
11.	THE HELL OF WAR	74
12.	MURCHIE STAGES A PRIVATE WAR	83
13.	THE BITTER RETREAT	92
14.	HORRIE SURVIVES SHIPWRECK	102
15.	THE WOG-DOG TO THE RESCUE	110
16.	WE LOSE MURCHIE	120
17.	HORRIE FALLS IN LOVE	128
18.	THE HAZARDS OF PEACE	137
19.	IMSHI TO THE RESUE	144
20.	THE WOG-DOG ATTACKS THE WOLF	154
21.	THE WOG-DOG IN DANGER	163
22.	HORRIE IN PERIL	172
23.	HORRIE TAKES IT ON THE CHIN	181
24.	THE STOWAWAY	191
25.	THE AUSSIE DOG	201

1

INTRODUCING THE DESERT, THE PUP, AND THE REBELS

A LAND of the dead. An oasis vanishing in the breathless heat. Sands of Egypt more pitiless still by reflection from parched black rocks. This is the Western Desert, the eerie land that was. On the edge of this desolation was our military camp, in the Ikingi Mariut area.

A phantom land. For that distant eastern city had long since vanished as the mirage was vanishing now, leaving bare the black range. Among those rocks were innumerable caves, the sepulchres of thousands of mummies. Truly, death had long since conquered this land which once built proud cities pulsing to the activities and voices and feet of men, women and children. I wondered what those folk had looked like. Fantastic now to imagine this desert was once a land flowing with milk and honey. But such was an historical fact in the pages of time.

Nothing, surely, could ever now live in this desolation.

With relief I dismounted from the cycle, wiped a sweating brow. Which introduces you to me, J. B. Moody of the 6th Division, A.I.F., training in Egypt for whatever the future might hold. A toss-up, I knew. Heads to join the sleepers of the desert, tails to see dear old Australia once again.

Somewhere in the distance behind, Don my mate was clinging to his motor cycle, bumping over rock to squish into patches of shifty sand. With eyes staring straight ahead he would be clinging to his machine, nursing it to the engine's limit, using all the desert cunning he knew while seeking to learn more, more.

We were playing a game, a game of approaching war. The knowledge gained was destined to save our lives when the armies of death came sweeping over the land. I pencilled a compass course on a slip of paper and buried it under a tiny clump of stones. I mounted again and rode away on that course, never deviating.

Don's job was to follow me and locate that heap of rock, then follow on in the compass course.

A difficult job, to find the hidden course then ride straight as an arrow across broken desert, locate the course again, then speed miles farther in a different direction as indicated by the hidden instructions.

When commencing our game Don would allow me a start of fifteen minutes from the "Wog's" bakehouse, a rough shed where the bread and sand were baked for the troops by Arabs under the supervision of Tommy soldiers. Don would then follow on, catch me if he could; if not, he would locate me at the end of the final stage.

This morning I was now on the last stage. I chugged along to the end and dismounted. Propping the cycle up, I sat in the ridiculous shade it offered and dreamily lit a cigarette, awaiting Don.

Don Gill, with his steady brown eyes and quiet smile, dark hair brushed any old way, slight of physique but wiry as whipcord, Don who would stand by a mate to the last. We were attached to the Signal Platoon as dispatch riders.

Hearing the sound of the approaching cycle my attention was attracted by a whitish object. In surprise I saw it - a small pup racing from rock to rock, a grim earnestness in his obviously tiring movements. He poked his small nose under rock after rock, striving with might and main to lever up the impossible weight, only to dash away to another rock. As he panted past I fancied his eyes held the glare of despair. Not even noticing my presence, his straining attention was directed to the lizards he was so futilely chasing.

"Poor little beggar," I thought, "thin as a scarecrow, he's desperately hungry.

"What is it?" inquired Don as he dismounted.

"A funny little white pup; looks as if he hasn't had a feed for days."

"A pup! Away out here! Where on earth could he have come from?"

"Perhaps he's been abandoned by some Italian family who feared the bombing at Alexandria."

"He's just about knocked up."

"Yes."

"Do you reckon he's an Arab dog?" I asked doubtfully.

"Could be anything from the look of him," replied Don. "A foreigner for certain. Doesn't understand our language."

"Here pup, good dog, here boy!" cajoled Don.

He ignored us for a while; then as if for the first time attracted by human voices, he stood and pantingly surveyed us, quaintly defiant and suspicious.

"What a comical little joker," laughed Don.

He was funny. His coat was a dusty white emphasized by a sandy-coloured stripe running along his back. On quaint, stubby legs he stood barely a foot high. The front

legs were bowed like those of a miniature bull-dog, his long body out of proportion to his height. His extraordinarily intelligent little face was pinched and forlorn, with an expression now changing from dire suspicion to one of inquiring hope. His stub end of a tail rose erect; his sharp little ears alternately stood to attention then dropped at ease.

"He's doubtful about us," said Don. "He's not sure whether he can trust us."

"He certainly looks like he feels. I suppose the poor little fellow has been chivvied from pillar to post."

To our sympathetic voices the outcast's tail wagged invitingly. Then he regarded us in expressive imitation of a question mark.

"He thinks we might help him," grinned Don. "Come here puppy, old boy."

The pup answered with a knowing leer.

"He's not to be won with salt on his tail," smiled Don. "We'll have to win his confidence somehow."

We advanced towards him, cracking our fingers. He stood his ground.

"He's frowning," laughed Don. "Won't give an inch of ground if he can help it."

Then I developed a brain wave.

"Let's chase lizards with him," I suggested.

I gave him a hand by removing a rock under which a particularly fat lizard had evaded him. His little stub wagged furiously; he charged in to the kill only to finish up with a mouthful of sand while the lizard darted under another stone. The pup wheeled around with a yelp of frustration, gazed up with such an air of "Now, wouldn't it?" that I hadn't the heart to laugh at him.

We toed over another stone, and as a lizard scuttled away the pup was after it with an excited yap. Again and again his little stub tail waggled thanks for helping him, he chased lizard after lizard but all escaped him. He gazed up with irresistible brown eyes appealing for further assistance.

"Please take this seriously," he seemed to say. "I'm very hungry."

He would let us touch him now.

I could feel his little ribs were only just covered by his silky, short-haired coat.

"He doesn't seem to have been doing well," I remarked.

"Poor little pup," sympathized Don, and patted his head. "You're a little outcast and far from home. You've got no home at all now."

We gained his complete confidence. Wearily he rested his chin on my arm and closed his eyes.

"What are you going to do with him?" asked Don doubtfully.

But I knew he knew what we would do.

"You'll have to be jolly careful," cautioned Don. "Pets are frowned on, while the strict rule is that no pets are to be allowed to the troops when we march from camp."

"How many rules have the Rebels broken?" I asked.

Don grinned.

It was a problem to get him back to camp. "Impossible to ride over this country one hand and hold the dog in the other," advised Don. "You'd better leave your machine and ride pillion."

It was a rough trip back to camp, but the little dog appeared quite contented. We got him to the outskirts of the camp, hid him, then doubled back for my cycle.

2

THE WOG-DOG ENLISTS

WE returned innocently to camp about midday with the warm little dog smuggled in my shirt. At our tent all the Rebels were "bashing the spine", sprawled out in various attitudes of "I don't care".

"What on earth have you got there?" exclaimed the Gogg as he stared towards my distended waist.

"Apparently a midwife is needed," grinned Feathers.

"Introducing our mascot!" declared Don with a haughty sweep of the arm.

I produced the pup. He was an immediate success; he said plainly "Glad to meet you all". He stood on his ridiculous legs in the centre of the tent, his tail wagging furiously, a quaint grin about his little open mouth, his appealing brown eyes surveying all.

"Where on earth did you get him?" asked Murchie.

"Out in the desert."

"He's asking to be taken on the strength," laughed Gordie.

It was a foregone conclusion; nobody could resist the pup. We gathered around him, discussing the problem in dark conspiratorial whispers. Would it be possible to keep him when eventually we must march out off camp and yet evade the rule of "No

Pets"? Our own officer was a very decent sort but he might not care to connive at the breaking of a strict rule. We felt confident of our sergeant however, old "Poppa the Sarge". He was human as well as a soldier.

"It will all work itself out," said Fitz hopefully. "We'll solve each setback just as it comes along."

Enthusiastically we agreed to adopt the pup as our battalion mascot, to care for him with might and main and crafty subterfuge and, if necessary, defy the powers that be.

"The first job is to feed him," declared Feathers. "He needs it."

"We'll each smuggle across a little from the cookhouse," suggested Gordie, "when the bugle blows. The next thing now is a name. "

This was a poser. He looked so comical we could think of nothing at the moment to fit him.

"Call him 'Longfeller'," suggested Feathers.

The name fitted the pup with his quaint long body on little stumpy legs. But somehow his intelligent little face suggested something better.

Just then our officer, "Big Jim", stepped into the tent. He stared down at the funny little thing gazing up at him, with a wag of tail stump.

"What on earth is it?" he exclaimed.

"We found him out in the desert, sir," explained Don

Big Jim leaned down, patted the little pup and took him up in his arms. The pup immediately tried to kiss him.

"What name this feller dog?" grinned Big Jim. And we knew "Big Jim" would be all right.

"A quaint little fellow," he said; then gave a quiet order and walked away to his own quarters.

"We can't walk about calling out 'Longfeller, Longfeller'!" declared the Gogg. "The camp would think we were 'desert happy'."

"How about George?" suggested someone.

"George, my foot!" protested Gordie. "You'd have all the Wogs in Egypt answering the call." (Our name for the native Bedouin was "Wog".)

"Call him 'Roy' after good old Poppa," suggested Don.

"Won't do," declared Murchie. "It wouldn't be fair to the pup, he looks so intelligent."

"Could we get at his name through his breed?" suggested Murchie. "Anyway, what breed is he?"

"Nothing on earth," laughed Feathers.

"He must possess a family history of some sort," smiled Fitz.

"He couldn't be an Arab dog," said the Gogg. "He's too well bred."

"That's it!" declared Fitz. "Arab dog — Wog-dog!"

We all laughed. "Wog-dog" really did seem to suit the funny little fellow.

"Wog" was the Aussie soldiers' nickname for the Arabs. The Arabic meaning of Wog is "worthy Oriental gentleman." But the Aussie soldiers' interpretation was something very different as you would readily understand if you knew the type of Arabs that pestered us.

"We cannot insult him by such a name," declared the Gogg. "How about adding 'Horrie', a good old Aussie name. 'Horrie the Wog-dog'."

It fitted the pup like a glove.

"Mess parade — headquarters!" came a yell from the orderly N.C.O. outside the tent. We tied the pup to the tent post and wandered across to dinner, from which we soon returned each with a scrap of meat for the pup.

We put the collection on an old plate and confidently offered Horrie his dinner — a huge dinner. To our surprise he inquiringly sniffed here and there around the plate, one little ear cocked in quaintly surprised interrogation. Then he got to work, turned his tail to the plate and showered it with sand, occasionally turning round to use his snout as a shovel and poke the meat well down.

Soon he had completely buried the meat. He gave a final sniff at the little mound and then, satisfied there was no smell left, gazed up at us with a "What's next?" expression.

"Now wouldn't that rip yer?" drawled Murchie.

We couldn't understand it; the pup was obviously ravenously hungry.

"Perhaps the meat is too much on the bugle (smelly)," suggested Fitz doubtfully.

"You bet it is— as usual," said the Gogg. "We cannot blame the dog; I admire his sense. What's supposed to be good enough for soldiers isn't good enough for him. I wonder if he could be an 'Eyetie' dog," he added, "an Italian dog. That's it!" he said excitedly. "He's used to olive oil on his food."

"I'll scrounge a little from the R.S.P.," I volunteered. Meanwhile Don dug up the meat and washed it clean of sand. We sprayed it with olive oil and offered it to Horrie.

Did he eat? He fairly hogged into that meat; it vanished. We felt quite proud at having solved the riddle of the olive oil.

"A dinkum brain-wave," said Fitz admiringly. The Gogg bowed in grave acknowledgment.

"However did you work that one out, Gogg?" inquired Don.

"Just fluked it," answered the Gogg modestly. But he was secretly pleased.

"You blokes on your feet?" inquired a grim voice as Poppa stepped into the tent. "Hullo, what's going on here!"

"Meet Horrie the Wog-dog," introduced Gordie with a gesture.

"Don't go near him," cautioned Murchie.

"Why?" asked Poppa.

"Because he's free of fleas so far."

"Oh yeh!" replied Poppa. "Well he's among a lot of then now, anyway. You funny little sausage. Are you going to be our mascot?" And the Wog-dog gazed up at him with his tail plainly wagging "Yes".

"Murchie's snakes will have to go, though," said Feathers.

The first setback. Murchie was a wizard with snakes; he kept a box full of them under his bunk.

"I'll get rid of them," promised Murchie surprisingly. "There's no damn kick in them, anyway." And we sighed our relief.

"He's a bit lousy," Murchie remarked towards the pup.

"So are you," admonished Poppa.

"All dogs have fleas," quoted Gordie and scratched himself lingeringly.

"We'll give him a bath," suggested Don. "Scrub the desert off him. He must be a respectable soldier-dog from now on."

"How can he become a soldier-dog," demanded Fitz, "if you take his fleas away."

"Some soldiers I know," said Poppa quietly, "would feel lonely without them."

"You'll never be short of company, anyway," replied Fitz.

Just then Don returned with a tin and hot water from the cookhouse. We all gathered round and dumped the little dog into the tin.

"Now then, you blokes," said Sergeant Poppa ominously, "the sarn-major sent me here to detail men for a working job. How about it?"

In the sudden buzz of conversation no one seemed to have noticed the request for labour. It was uncanny how all hands vanished, leaving Don and me and the pup.

"Can you beat that!" declared Poppa wrathfully as he jumped up and strode for the tent door. "Call themselves soldiers! They're not soldiers' bootlaces!"

Don laughed at Poppa's retreating footsteps as he strode out to hunt for the Rebels.

"He's got as much chance of picking any of them up as I have of hopping to Cairo," grinned Don.

"Here's my chance to pinch Fitz's soap," I said.

3

THE REBELS

HORRIE looked very miserable in his bath; he'd long since forgotten what bath-water was. But he suffered the indignity in disdain at our astonishment at the army of fleas a small pup can harbour. We dried him on Fitz's towel, which pleased him more than it would Fitz. Happy as Larry, he scampered in and out of the tent and up and down upon the bunks making himself at home.

"'He'll stay," declared Don.

"He's joined up with the 2nd A.I.F. for the duration," I replied, and felt really pleased.

"Better not make too sure," warned Don. And I knew he was thinking of the day when we must get our marching orders.

In time for the sundown meal the absconding Rebels came innocently strolling back to camp. It gives a soldier a glow to dodge a fatigue.

Horrie the Wog-dog enjoyed his evening meal then waddled thoughtfully to my bunk, his bingy almost touching the ground. He let himself flop to the sandy floor then gazed at us with one ear cocked, saying plainly as words "Thanks, boys. I've had a great day."

Don and I scrounged a few boards and some straw from the quartermaster's store. We rigged Horrie's bunk comfortably in the middle of the tent while the boys stripped off and sprawled at ease in their bunks, lazily smoking as they watched the show. The Wog-dog heaved himself up and waddled across to see what was doing. Sitting back comically he watched proceedings with keen interest, his head cocked to listen as the lads passed some remarks.

"That pot-bellied, shrewd little desert rat knows every word we're saying."

And the Wog-dog glanced cunningly at the speaker.

"He knows the bunk is being built for him," declared Fitz.

Don and I tried to induce him to enter the bunk, but he developed another interest in life and sidled towards Gordie's bunk with speculative expression. Gordie snatched down his socks as the pup sprang to grab them. With an excited yelp he wheeled around and charged Fitz's bunk, but Fitz was just in time. Feathers was a second too slow, for with a delighted yelp the Wog-dog ripped the socks from his hand and made straight out the tent door.

At last we coaxed him into his bunk from which his little head gazed at us saying: "Very well, now you've got me in the box! What's the fun now?"

He answered his own question by leaping out of the box and diving under Murchie's bunk. In an instant pandemonium broke out, frantic yelping of the Wog-dog, showers of sand.

"Your wretched snakes!" yelled Don and dived under to rescue the Wog-dog while Feathers laughed delightedly. Feathers could see fun in the most serious situation.

By the time Don dragged the excited pup away inquisitive heads were poking in through the door of the tent.

"Take your wretched snakes to the devil out of this!" Shouted Poppa.

"Go fry your face," snapped Murchie. "I'm going to try an experiment first; see if I can get any life out of them."

"Your beastly snakes will kill a man one of these days," roared Poppa. "If ever you bring another pet snake into my sight, I'll dong it and you too!"

"Go and bag your head," growled Murchie as he dragged the box from under the bed. Lifting the heavy stone from the lid he gazed fondly down upon his pets. And a nasty lot of wrigglers they looked.

"Get your experiment over," demanded Don as he clung to the Wog-dog.

With a fanatical gleam in his eye Murchie thoughtfully began preparing some apparatus consisting of several copper wires, a gadget or two, and a car battery. We watched while the struggling Wog-dog barked himself nearly into hysterics.

"Quieten that pup," commanded Murchie as he fondled an asp, "so that I can explain the experiment. It's this way," he went on and slowly connected a terminal to the battery. "Cleopatra died by clutching an asp to her breast after learning that Mark Antony had been killed. Now, Cleo had it and plus — no man was safe within a hundred miles of her. I've often wondered what this fatal attraction of hers was — whether it had something material in it, something like magnetic waves or personal electricity or something like that. If I could find out what these 'Come-hither' waves were, then extract 'em from the air or corpses or something, if only I could bottle up this 'Come-hither elixir' I'd make a fortune after the war. Pondering over this and trying to analyse Cleo's feelings I've come to the conclusion that the secret lies in her vibrations at the time. She must have been all keyed up. So I wondered how did she grasp that asp. Did she fold it to her breast with gently trembling hand, or did she crush it to her like a centurion in a death grip? I'm going to find out."

He picked up the other lead from the battery and stabbed an asp with a charge of electricity.

Murchie went sprawling, Rebels leapt up to bunks and a frantic Wog-dog was chasing agitated snakes all across the floor. From outside the tent a scurry of sounds told of a lively stampede.

When it was all over Murchie proceeded to crawl under bunks, searching for snakes. Every now and then from a tent outside would come a yell followed by sounds of evacuation. Sounds arose of an angry Protest Committee approaching the tent.

Thus passed the Wog-dog's first night in camp; it was hopeless to try and hide him now.

Well, you've met Horrie the Wog-dog, so I'll introduce you to ourselves.

Officially, we were the Signal Platoon of the 2/1 M/G. Battalion. But to the camp we were "The Rebels". Always one, and usually two or three were in trouble, through no fault of our own.

"Who's for the 'mat' today?" would generally be Feathers's lazy salute to the dawn. And sure as fate some gloomy soul would have to toe the mat, or otherwise dodge retribution. On many a morning the sergeant's report of our particular Section on Parade would read:

"On parade, 10. Duties, 5. Sick, 2. A.W.L. 2." We would try all we knew by innocence and bland subterfuge to camouflage the misdeeds of our mates away without leave, and often managed to lay a smoke-screen between them and wrathful authority. You see, there was the fascinating Lost City not so very far away to explore, a dead city of ancient times, while not so many miles away rose the eastern spires of a very attractive, very wicked live city — Alexandria.

George Murchison was "Murchie" to us, a venturesome young Aussie with a ready grin. Mischief and he were cobbers; he was our Signal Officer's No. 1 "Problem Child". His sandy hair was seldom brushed, his boots were dusty, his smiling cheeks decorated with patches of stubble. His manner of shaving was to "get the jolly thing over." He'd tear the razor once down each cheek, then down the chin and uppercut the neck. He'd murmur "Thank goodness that's done!" and wipe the soap off on anybody's towel. An absolute menace with firearms, he had pock-marked the tent roof with bullet holes. To our startled yell he would grin "Sorry! Didn't know she was loaded." We were suspicious of these "accidental" shootings for Murchie was a crack shot and an expert with rifle mechanism. He would clean the lethal weapon with a care ludicrously at variance with his care of his own person. Absorbed in his task when we were immersed in our own, suddenly "Bang!" and another bullet sped through the roof.

"You flaming coot!" Poppa would roar, "you did that on purpose!"

"Sorry," Murchie would grin. "She just went off!"

"And so will you one of these days, you grinning hyena," Poppa would declare.

Fate had burdened Murchie with an insatiable love of adventure and two nerve-racking kinks of humour — his joy at frightening six month's growth out of us with an unexpected rifle shot, and his mania for collecting snakes. He would wander for hours in the desert, overturning stones and collecting asps and other poisonous crawlers. These he secreted in a box under his bunk and waited quietly smoking in the evening for the unearthly yell when some unlucky wight trod on a reptile surreptitiously released from the box. Murchie really was trying at times, but a better mate never wore boot-leather. As to soldiering on, he was always hungry for action and was an excellent signaller; this helped our officer, who was keen on efficiency, to overlook Murchie's shortcomings. You see, there were no snakes in the officer's tent.

George Harlor was "Gordie" to us — a lean, dark, brown-eyed quiet chap not so quiet under the surface. His particular hobby was making wireless sets from bits of wires and gadgets he'd souvenired here and there. At most inopportune moments our conversation would be interrupted by ear-splitting shrieks and whistles as Gordie tried out some new machine. Sometimes the shrieking gadget would "run away from him", as Poppa expressed it, while we held our hands to our ears and yelled in angry opposition, which did not seem to worry Gordie in the least. He was to develop into a wizard in wireless but his experiments certainly frayed our nerves. Very popular with the Rebels, he'd be in on anything and could always be relied upon to think of a quick way out when we got into a scrape.

Bert Fitzsimmons "Fitz" to us, was a six footer. Blessed by a pair of laughing blue eyes he was wonderfully quick-witted, a master at repartee; whether resting in the tent or on the march he could keep us alive with his sparkling wit. He and Gordie were nearly inseparable. Fitz loved a joke and was a dinkum soldier too. His pastime when nothing was doing was the great national game, "Two-up". In leisure hours Fitz could generally be found at a two-up game. When the officers declared a blitz against two-up, Fitz's quick wit countered by organizing "Marbles". The ring was drawn as close as possible to the officers' mess, and these gentlemen stared in puzzled amazement at their harum-scarum lads deeply immersed in the game of marbles. The marbles were a camouflage for pennies.

Brian Featherstone was "Feathers" to the Rebels. The baby of the section, he was a tall lad with twinkling blue eyes, a mop of sandy hair and a fresh, boyish complexion. The direct opposite to Murchie in general appearance, he was always smart in attire and on parade. An excellent signaller, he earned stripes but his irrepressible spirits made the holding of them highly problematical.

Delighting in a joke, he saw fun in everything and was the bane of Poppa's life; he was in his element when slyly teasing Poppa. That grizzled old warrior darkly hinted at the dire happenings Fate held in store for the larrikin. A lot of Poppa's time and persuasive language were spent in getting that same larrikin out of innumerable scrapes.

Bill Shegog was "The Gogg". The Gogg was a tall chap with medium-coloured hair and somewhat sallow complexion, a signaller as good as the best of us and with a dry wit of his own. He was a "hard doer" and one of the "characters" of the section. Among his other worthy attributes the Gogg was blessed with that priceless gift of "smelling out" beer, should any of that delectable fluid be secreted within miles of the camp. A keen sportsman, he was also an artist with brush and pencil; he brightened us up with sketches of dusky and unclothed damsels. The Gogg was also in on anything that was doing, and did not mind mixing it in a brawl when any of the section got into trouble.

My particular cobber was Don Gill, my dispatch-rider mate. A rather quiet chap with very dark hair and brown eyes, Don was destined with me to run the gauntlet in many a breathless ride in the grim days swiftly coming. A good soldier, though a bit of a wag, he would stick like glue and was very popular with the section.

And now we come to "Poppa", grizzled old veteran Roy Brooker, the section's signal-sergeant. We were very lucky; Poppa was a real gem even though a rough diamond.

He had apparently been somewhat absent-minded when giving his age on enlistment. This was a so-called "young man's war", but Poppa must have seen at least fifty summers and winters go by. "Mostly winters," declared Fitz, "judging by that weather-beaten countenance."

"It is a face that can withstand the ravages of time," Poppa would reply, "not like some faded roses I could name."

Poppa's mousey hair was not too plentiful, but service with the 1st A.I.F. at Gallipoli and in the Desert had not cured his appetite for adventure nor slowed him up at all. He often set a pace that made us younger fellows sit up and take notice. A Dinkum Aussie was Sergeant Poppa — one of the best. Secretly, the Rebel section was the pride of his life, though he'd never admit it. The old tyrant would roar threats of a certain court martial at us each time we were caught in another wild escapade. Still rumbling threats he'd stride out of the tent en route to headquarters, and when he got there he'd work every old-soldier trick he knew to get us out of the trouble.

Yes, Poppa was the strict disciplinarian with the heart of a lion, a kindly old lion. It took us a very long time to realize that he possessed that priceless knack of understanding and sympathising with anyone's hidden human complexes, without the man concerned having the faintest idea. With this priceless instinct in his make-up it is easy to understand that Poppa was able to get the very best out of any man, and that best was given willingly.

And now, our officer. Last maybe, but no one would have said he was the least. Lieutenant Jim Hewitt, "Big Jim" to us. A wonderful physical specimen of over six feet, with shoulders like a working bullock, and a capacity for work that would have made a draught horse appear like a sissy. A keen and efficient soldier, his sense of duty we secretly guessed sometimes clashed with his sense of the human. He would do anything for us but he demanded and got the best from every man. With an officer like him and a sergeant like Poppa, the Australian private is unbeatable by any troops in the world, no matter how harum-scarum, irresponsible headaches those same privates may be when there "isn't any war on". Big Jim was a great sportsman also, and particularly keen on anything he tackled, whether military matters, sport, or anything else. There were times, indeed, when we strained his patience and it would have gone hard with us were not each man thoroughly efficient at his job. Even when we broke over the traces though, we had his complete confidence and in return would have backed him up to the last. This mutual wariness but confidence was to stand the Wog-dog in good stead for we suspected Big Jim turned the "blind eye" on various coming occasions when it was touch and go as to whether the Rebels were to lose the Wog-dog or not.

And so you know all of us of the Rebels. And history tells you all about our battalion. Now to the adventures of Horrie the Wog-dog.

4

MURCHIE'S ASP

The battalion warmed to the Wog-dog; despite his sock-tearing pranks he became a great favourite. His appetite amazed us. He always beat the bugle and ushered in meal-times by a swift waddling from the tent to the cookhouse. Even so, he was first on parade too, especially the route marches. He marched proudly at the head of the column, his quaint little legs stepping out with the funniest imitation of a martial air. Head and stumpy tail erect, chest forward, ears pricked, his keen eyes invited all and sundry to "Step out! Step out!"

During a route march the Rebel section manoeuvred to be in the leading platoon and keep an eye on the self-important Horrie. As he waddled along at the head of the column he believed that everything should give him right of way. But we were afraid of the busy trucks constantly tearing past. Horrie quickly learned to obey the Rebels but showed rebellious tendencies should any other soldier order him about. I first trained him to obey commands by aid of a fishing-line tied to his collar, while he marched proudly ahead with a glance back at us for approval. Now and again I would order "Come alongside!" then pull in the line until he came to heel. Very soon, at the order "Come alongside!" from any one of the Rebels he would immediately obey. We soon discarded the training string; his aptitude in grasping the meaning of anything required of him was uncanny.

"He is the most intelligent member of the Rebel section," growled Sergeant Poppa, "and," he added as an after-thought, "certainly the most obedient."

"That's because he's learning all the old-soldier tricks," laughed Fitz.

"What you don't know, you cub," growled Sergeant Poppa, "would fill a book — a library of books."

Horrie detested natives. Probably they had been cruel to him during his homeless days before he took to the desert. Where we were camped on the fringe of the Western Desert there was nothing but our camp, sand and barren rock, and the occasional humble camp of the wandering Bedouin. These extremely poor but surprisingly proud children of the desert were not disliked by the Australian soldiers as were the breeds that swarmed in the cities and their outskirts. We had to keep a sharp eye against the Bedouin on account of his expert thieving propensities but apart from these, in this particular camp he never caused us any trouble. The Bedouin children were rather favourites, and very different to the cheeky, baksheesh-cadging urchins of the city native quarters. But all native urchins were the same to Horrie; fiercely he growled his protest when we ordered him not to chase them. He would fly at any grown Arab who approached within smelling distance of the camp. This made it safe for us to leave washing to dry on the tent ropes overnight, so long as the garments were out of Horne's own reach. Prowling shades of the night soon learned to avoid our tent.

During the long route marches we would gratefully obey the occasional order to halt and smoke-oh. At such times Arab urchins would spring up from nowhere, vociferously selling us "eggs acook!" water-melons, sweetmeats and native lemonade, though with a wary ear on Horrie's growls of disapproval.

These quick-witted little raggamuffins were shrewdly intelligent. A common trick of theirs was to say deferentially: "Gib it baksheesh, sergeant!" knowing that the "sergeant" was but a humble private.

They were good-humoured, surprisingly quick to grasp Australian slang phrases, and smart in picking up odd items of information which might help them in selling their wares. They had lent on attentive ear to the rules of hygiene that necessarily were so constantly drummed into the troops. While treating our idea of cleanliness with contemptuous mirth, they earnestly claimed their goods as "Verry clean!" "Verry sweet!" "Verry hygiene!" as they offered the fly-specked goods with grubby little hands that had never known a wash.

One day a quick-witted little chap sold out his goods because of the roar of laughter he caused at Horrie's expense. With serious mien Horrie took advantage of this particular smoke-oh to make his toilet in view of all. As energetically he scratched sand over the result the urchin shouted "Very good! Very sanitary!" While another quick-witted youngster cashed in on the laughter by yelling "Now wouldn't it!"

Horrie had an idea that the laughter was at his expense and with a menacing growl turned on the urchins, and despite his size caused a lively scatter.

"Big Jim" led the platoon on a route march one day and as we approached a small Arab village he roared the order: "March to attention!"

All backs were straightened up, rifles sloped automatically, heads up, chins in, chests out to the rhythmic tramp of marching feet. Big Jim evidently wished to impress the natives, for as a rule we marched at ease and yarned and walked along pretty well anyhow.

Dozens of Arab urchins flocked from the dwellings to gaze at the impressive sight; we were putting on a brave show with the Wog-dog, tail erect, marching in the lead. When right in the centre of the village a tiny urchin shrilled: "March at ease!" and the splendid sight collapsed to the laughter of the boys.

"Saieda (good day) George," called Murchie to the lad. His sharp little eyes took in Murchie's usual disreputable appearance and instantly he shrieked "Good day, Wog!"

Murchie was furious, the troops delighted. Any man could have bought a fight with Murchie if he had been game to say "Saieda, Wog!"

5

ONE NIGHT IN ALEXANDRIA

HORRIE's education went on apace; he was such an eager pupil that the lads took pleasure in teaching him.

His lively interest in everything was another trait in his character which endeared him to us. Often when the Rebels were discussing some matter he would squat quietly there gazing up at our expressions and catching every word. Particularly when during Sergeant Poppa's absence we would be plotting some dark deed against Authority, Horrie's big brown eyes would fairly dance with delighted comprehension. Always eager to join in our escapades we had but to whisper "Be quiet!" and he would move silently as a mouse.

He was fussily pleased with his little box bunk and during the hot hours would recline in it while keeping one eye cocked towards the tent door. Nothing delighted him so much as to get in first with a greeting to any Rebel entering the tent, or to rush out if he believed someone was threatening the Rebel's possessions.

Nights in Egypt were surprisingly cold and Horrie discovered that the foot of my bed was warmer than his box. Shortly after lights out, when our conversation had ceased he would snuggle on the bed but the slightest movement would send him scuttling back to his box until I settled down again. Sometimes I'd strike a light and frown, and there was his little head poking from the edge of the box, innocence in the big brown eyes.

He loved to accompany the inspection officer through the ranks during inspection parades, standing by with critical eye and knowing head as if detecting as keenly as the officer any shortcomings in the men's kits or untidiness among the tents. After inspection, he would take up his position in front of the platoon facing the troops, for except on marches he would not turn his back to the ranks.

He was humanly downhearted by occasionally missing a parade. At the sound of marching feet he would rush out barking for us to hurry and join in. But when we did not come he would return on his own account quite crestfallen though understanding it was not the Rebel's platoon that was marching.

"Next time, Horrie," I'd promise and he'd wag his tail with brightening countenance while enviously gazing after the platoon.

One morning we received orders for a battalion parade. Our company formed up in platoons preparatory to a preliminary inspection before moving off to the battalion parade-ground. Corporal Feathers happened this morning to be in charge of our signal platoon. As usual he was spick and span as a new pin, the smart, perfectly equipped soldier, erect at the head of the platoon. Horrie was yelping excitedly, racing round and urging the platoon to move off. Just as Big Jim came striding up to take command of our platoon, another came marching by and Horrie's excitement burst control. In the desert there are no trees where a dog can express his feelings but Horrie's anguished eye lit on Feathers's beautifully polished boots. In a second he had dived up behind the immaculate soldier and cocking his leg, made his salute in full view of the platoon just as Feathers smartly saluted Big Jim.

Even then the platoon might not have exploded had not a stray mongrel noticing Horrie's mistake dashed up and added his quota. The platoon burst into an uncontrollable roar of laughter. Feathers quite unconscious of the comedy turned about sharply and ordered "Quiet everywhere! Every man firm as a rock!"

Fitz called out, "That's what Horrie thought, sir!" and it took ten minutes and a threat of seven days G.B. to quieten the platoon, all the more tickled by the puzzled frown of Big Jim who could not make out what had happened to his platoon on parade. Horrie meanwhile dashed to the head of the platoon impatiently wagging his tail while awaiting the order to march.

The little dog dearly liked a trip in a military truck and would stand by the side of one, asking for a ride. It was a good adventure for him to stand on the driver's seat, bark at the Arabs and assault the dogs of Alexandria.

By the way, it was in Alexandria that Murchie blotted the not-so-lily-white escutcheon of the Rebels. I won't say we were on leave, exactly, but anyway we were there. Night. Under the stars the massed lights of Alexandria blinked upon the historic city.

We strolled into a cafe where the red-fezzed orchestra was fairly drowned in the roar of voices; the lights were dim through clouds of tobacco smoke. New Zealanders, South Africans, British Tommies, Free French, Aussies, Allied soldiers of many nationalities were all at their tables in lively enjoyment of precious hours of leave. As we sat down and ordered our drinks I handed my hat to a waiter.

"Hang this up too!" said Murchie and handed him something. A scream silenced the cafe as the waiter leaped back in ghastly fear from the asp he had dropped.

Murchie's little joke. The asp wriggled and the waiter screamed again and fell backwards over a table. That did it; there was a crash of glasses and foam of spilt beer, and the Tommies picked themselves up with howls of anguish. If you wish to arouse the British Lion just knock over his beer. They flew at us and a brawl was in full swing. Tommies rushed to join Tommies, Aussies sprang to join Aussies while Allied soldiers sprang up in noisy gesticulation. Overturned tables, upended waiters, a frantic orchestra, howls from the management while a cosmopolitan mob gaped in from the street. Then concerted yells of "Coppers, coppers!" broke up the struggling groups as all fought for the exits to escape the military police.

"Not a bad start," panted Murchie as we picked our-selves up out of the gutter. "We've only been in Alex twenty minutes!"

But a much more disturbing event happened back at camp. When nothing much was doing Horrie was in the habit of wandering through the camp, visiting his numerous friends in their tents. One day he returned to us limping. We examined the leg that was worrying him but could find nothing wrong. Later that day we were informed that Horrie had shot out of one particular tent with a yelp, the only occupant at that time being a particularly nasty piece of work who unfortunately was able to shelter under three stripes. We could not understand any man who would seek to hurt Horrie but this man was the only one in all that battalion who detested dogs.

"I'm going to do him in!" declared Murchie with an ugly look. He was striding towards the offending sergeant's tent, and would listen to none of us.

"Look here, Murchie," I protested, "for goodness sake don't thump him or we will lose Horrie. No dogs are allowed in camps, remember! We'll have to face a serious charge of assaulting a sergeant and we won't have a leg to stand on. It will mean we will have to give the Wog-dog up."

In frowning anger, Murchie hesitated; we talked it over and walked slowly back to the tent.

The time came when the particular insect who had hurt the dog happened to be sergeant of the guard. It was a pitch-black night. One duty of the sergeant was to visit each post to see that all was in order. While on this tour of inspection he was suddenly confronted by a muffled figure and challenging voice.

"Who goes there?"

"The sergeant of the guard!"
"Sergeant so-and-so?" inquired the figure.
"Yes."
"Cop this!" and the sergeant was knocked to the ground.

Next morning he sported a lovely pair of black eyes. Strenuous efforts were made to find the perpetrator of the dastardly deed, for the "knocking" of a sergeant of the guard was unheard of. A check-up of every tent was ordered but all the Rebels were in their tent at the time, Poppa, the signal-sergeant was there to prove it!

The offender was never brought to light despite the persevering attempts of the black-eyed sergeant. For some reason or other, he was certain one of the Rebels was the culprit.

But the dark crime faded into the limbo of unsolved mysteries.

6

DEEP LAID PLANS

THE wildfire the rumour spread "Something doing! There's a move on!"

No one knew how it originated, no one could obtain any verification or details. Any diplomatic approach to an officer brought a surprised lift of the eyebrows. No, he'd heard nothing about it. Or he would shrug the matter away as just another rumour.

But we noticed that the officers were taking a keen interest in life.

It was about this time that the Gogg got mixed up in an argument with a gharry in Alexandria. You cannot ignore a horse, gharry and wild Egyptian driver — not when they're galloping "all out". They took the Gogg to hospital where he bemoaned his luck in getting knocked just when a move was on. He declared he'd follow us up, even on crutches.

After many rumours, the actions of Sergeant Poppa first convinced us there really was something doing. In the temporary absence of Big Jim he ran our particular little show. His enthusiasm in checking the signal gear and in bullying us for mislaying it was a sure sign something was doing, particularly as he was so mysterious about it all.

"Like a clucky old hen hatching eggs," grinned Gordie.

"It was a mistake when you were hatched!" declared Poppa.

"A better job than you, anyway," came back Gordie. "You're only a bad egg."

But we could not draw Poppa though we saw he was dying to spill the beans.

"We know all about it, anyway," declared Don. "We're going home so you can lead the Anzac march through Melbourne."

"Too ancient," said Fitz pityingly. "He'd never see the distance out."

"Neither of you will ever lead any march," snapped Poppa. "When you're my age you'll be wheeled about in a bath chair."

"Poppa don't need his bath chair," remarked Gordie. "We've carried him a long way now and it looks as if we've to keep on carrying him."

"When you've got to carry me," said Poppa grimly, "it will be in a box."

Next thing was that Don and I were issued with two brand-new motor cycles; while running and adjusting the engines we eagerly discussed the certainty of a move. Wild guesses again swept the camp. "Greece?" "Western Desert?" But to the Rebels' inquiries Sergeant Poppa still kept a stern silence.

Making an excuse that we were testing the new bikes, Don and I slipped away to Alexandria. I wanted to buy some spools for the camera, anyway. We entered a Greek shop the proprietor of which spoke excellent English.

"Sorry, boys," he said, "no spools. But you'll soon be able to buy plenty in Greece!"

We walked out quietly; the Greek knew more than we.

As we rode back we both were thinking about the Wog-dog. He must come with us. But how?

"He's solved it!" declared Gordie when we told what the Greek had said. "The weight of our gear has been assessed and the boxes are measured in cubic feet. That points to a short sea trip where space will be vital."

Fitz voiced our thoughts.

"What about the Wog-dog?"

We were discussing the pros and cons when Poppa entered the tent. He went straight to the Wog-dog, picked him up, sat on the edge of my bunk and stroked him.

"Poor little Wog-dog," he said. "We'll be sorry to leave you."

"Where we go Horrie goes also," snapped Feathers.

Poppa put the dog down and grinned happily. "I was wondering if you were slipping. That's all," he answered.

"You prehistoric worm!" declared Murchie. "If we leave anyone behind it will be you!"

"Relax," grinned Poppa. "You'll get all the scrapping you want before long."

"So the move is on?" I inquired.

"It is," replied Poppa, and lapsed into silence.

"Are you going to stay behind to mind the kit-bags?" inquired Murchie sweetly.

"You'll take your own kit-bags," snapped Poppa, "and I'll take you!"

We gazed at him in respectful silence. At last his weathered countenance wrinkled into a grin. Sergeant Poppa had relented.

"Look here," he said confidentially, "this is the news — but keep it under your hat! First of all, mail closes tomorrow at 1700 hours. So get your letters to the orderly room before 5 o'clock. The transport is going ahead the following day and I suggest there might be room among their gear for something the size of a little Wog-dog. Don and Jim will probably take their machines and travel with the transport.

Good old Poppa. The first stage of the problem was solved by that cunning old head.

"I'll trot over to the transport tent and chat to Ron Ford," I said. Ron was one of the drivers. I found the transport platoon busily writing letters home.

"Fordy about?" I inquired.

"He's down at the truck lines," answered Reg Jenks. "Did you get the drill (news)?"

"Yes. I hear you pull out the day after tomorrow."

"Yes. Looks like Greece."

"That's the Rebels idea, too," I replied, and located — Ron Ford.

"We're in a spot, Ron," I explained, "about Horrie. There's a good chance we could smuggle him away if only you could fit him in among the gear in your truck."

"Of course," agreed Ron. "The battalion would never forgive you if you left the little dog behind. He can travel in the cabin of the truck and I'll fix the windows so that he can't be seen."

"Thanks, Ron." And we arranged for such details as a tin of water and doggy meat for several days.

"We'd better put in a fair issue of water and meat," advised Ron. "You see, the truck may be lowered down into the ship's hold and I may not be able to get at it for a few days."

"Phew!" I said. "Horrie will be terrified."

"We've got to take risks," explained Ron. "If the truck is lashed to the deck then of course it will be all right. But we don't know what will happen when we get to the wharf. I'll do my best."

"The sarge tells us he may be able to detail Don and me to travel with you. If so then one or the other of us will be able to watch where the truck is loaded aboard ship and rescue Horrie if anything goes wrong."

I returned to the tent and discussed the scheme with the Rebels. All agreed that it appeared quite practical so long as Ron the driver, Don and I could work together with an eye on Horrie.

Next morning came the first shock. Despite Poppa's pulling of strings, Don and I received orders to load our machines on one of the trucks and travel with the signal platoon — not the transport platoon.

The more we thought of it the blacker it looked. The Rebels held serious council.

"Ron Ford will do all he can for the little Wog-dog," said Gordie, "but we know what happens when troops move; a man may be transferred to a different job at any time."

"I hate to think of the little Wog-dog away down there in a dark hold locked alone in the cabin of a truck," said Fitz.

"We wouldn't forgive ourselves," said Murchie, "if the ship was torpedoed and we didn't get time to go down and give him his chance."

That decided it; we agreed we must straightway work out another scheme.

"We must get him aboard somehow," said Don. "Then he'll be all right."

"No, he won't," cautioned Fitz. "Quarantine regulations are too strict; we still must hide him or some of the crew might put him away."

"They wouldn't do that," said Gordie.

"Not in the main, they wouldn't, but some semi-officer bloke might spot him and it might be more than his job was worth to keep quiet."

"Get him aboard first," advised Gordie. "We'll solve the other problems as they come. Now, he's such a tiny little chap that we could easily hide him in one of our kit-bags."

"A brain-wave," declared Feathers.

"It sometimes affects me that way," said Gordie modestly.

"Yes," smiled Feathers, "though seldom in the right place. But let us work on this kit-bag idea."

"I wonder if he would keep quiet," I said. "If he should bark at a critical moment he would put the show away. We could hardly keep him in the bag for two or three days either. Don't forget he's got his own little arrangements to make."

"I've got the right answer to that problem," declared Feathers. "It's about two days travel from Alex. to Greece, isn't it?"

"About that, all going well."

"Well, counting Poppa, we know that six of us Rebels are detailed to travel together. We get the dog aboard somehow per kit-bag. Six of us will be together. Forty-eight hours is two days. Six into forty-eight goes eight."

"Marvellous!" declared Murchie.

"How well educated he's been," said Fitz admiringly.

"He went through college on a bike," explained Murchie.

"And thus," resumed Feathers, "we arrive at shifts of eight hours, if your limited intelligence can grasp that simple mathematical calculation. Now, once aboard the ship and we will scatter and search for a latrine. When located we claim it speedy like. The rest is childishly simple. That latrine will be Horrie's suite; in it he can live in comfort and attend to his toilet as and when required. One man will stay with him throughout each eight hours, each of the six taking turn about. Providing the caretaker has a book to read, the time will pass very pleasantly, the seating accommodation being all that is desired."

"Marvellous!" breathed Fitz.

"Scrumptious!" declared Gordie.

"I'm for the first shift!" claimed Feathers.

"Don't move so fast," warned Murchie. "Wait till the proper time."

"That'll be O.K." grinned Feathers. "I'm regular."

"It's a good idea, anyway," said Don. "I'm relieved."

"Not yet," grinned Feathers, "It's not your shift."

"Oh go and fry your face," replied Don. "The idea's good anyway."

"Perfect!" said Feathers with a wave of the hand. 'We can easily take his meal to the man sitting on shift. The scheme is perfect."

"Yes, Solomon," broke in Murchie scathingly. "Only for one little point you've missed."

"What's that?" demanded Feathers.

"The latrine may not be a single-seater!"

This was a blow, we knew only too well the usual arrangements made aboard ships for troops — a rolling line of benches with seating accommodation for a troop of men open to the world.

"There must be a single-seater somewhere aboard a ship," declared Feathers, "and we simply must find and commandeer it, even if it's the captain's."

"Well, we'll trust to luck about the single-seater," I suggested. "Let us first educate the Wog-dog to the kit-bag. I'm afraid he'll play up. By the way, where is he?"

"Poppa has him on duty guarding the signal gear to keep the Wogs away," answered Gordie.

"I'll go get him," volunteered Murchie, and strode to the tent door.

"Better speak to him as you approach," called Feathers, "or he'll mistake you for a Wog."

"That's why you didn't offer to get him," came back Murchie's voice. "He'd smell you a mile away."

Murchie returned with Horrie and Poppa.

"Fitz," ordered Poppa, "you go and take over Horrie's place for a while."

"Oh," growled Fitz, "it would be a poor show if any of your precious gear were pinched at this stage of the game!"

"It would," snapped Poppa. "Go and guard that gear. Step lively."

"Oh, all right," growled Fitz. "I suppose I've got to do what you can't do," and he growled his way out into the dark.

'Well, what's the new idea?" demanded Sergeant Poppa.

We told him. 'Well, well," he declared. "So my patient teaching is bearing fruit at last; I was beginning to think you were hopeless."

"You've got nothing to think with," protested Murchie.

"An extra fatigue to you for that crack," replied Poppa amiably. "And now, you poor nit-wits, I'll run the rule over your plan."

"You couldn't improve it," declared Feathers.

"Nit-wit," answered Poppa scathingly. "Paint on a card-board 'Out of Order' and tack it on the latrine door."

We saw the point at once.

"The dear old man," said Murchie admiringly. Who'd ha' thought it".

"Not you, anyway," replied Poppa.

"He was always such a stickler for privacy," grinned Murchie.

"I wasn't dragged up beside a hollow log," explained Poppa.

"There's one snag," murmured Gordie.

"Well, what is it?" snapped Poppa.

"If the latrine has 'Gentlemen' painted upon it then you won't be able to go in and do your shift."

"He'd ignore the notice," grinned Feathers, "not knowing the meaning of the word."

"Get to work and educate Horrie to that kit-bag," ordered Poppa.

I held the bag while Don introduced Horrie stern-first into it. Horrie kicked a bit but when I lowered the bag to the floor and his little hind legs found solid ground he ceased kicking, standing on tip toes in a comically desperate attempt to keep his head out of the bag.

'We can't shut the bag," said Gordie doubtfully, "for if we pull the neck-string tight Horrie would smother. That means that we cannot carry the bag the usual way."

"Cut an air-hole in the side of the bag," growled Poppa.

We did, and poked Horrie's bewildered head out of the hole.

'We can carry the bag now as usual and Horrie won't choke," said Fitz proudly.

Poor Horrie looked so comical we all laughed; gratefully he gazed up at Poppa as the old warrior patted his head.

"Never mind, Horrie; you'll soon take a tumble; it's you will soon be laughing at the nit-wits here."

When we let Horrie out of the bag he shuffled shame-facedly to my bunk and, settling down, rested his chin on his leg and looked very miserable.

"The little chap is offended because we laughed at him," I said as I picked him up.

"Pretty smart dog, our Horrie," flattered Poppa and soon under the reassuring words the little stubby tail was wagging again.

At the second trial he entered the bag more willingly though not actually liking it, but he seemed resolved it was part of some new game he must play.

We saw he would quickly learn every trick we could teach him. We considered the problem of getting him on board ship unobserved was now as good as settled. When once aboard, the locating of his "suite" must be left to quick wits and fate.

7

THE WOG-DOG EMBARKS
FOR THE FRONT

NEXT day the big camp was busy packing up for the move. All knew that battle, hardship, sudden death for some, all the terrible chances of war, lay ahead. In many a man's mind was the rumble of guns far away.

Having put the final touches to our signal gear, the Rebels carefully packed their personal belongings. My gear was divided amongst the Rebels, for mine would be the full-time job of carrying Horrie in the kit-bag. With a quaint earnestness he watched our operations; he sensed this was no play, no preparing for any ordinary route march, but that something serious was afoot.

Long after lights-out that night I lay awake; Horrie crept from the foot of my bunk and rested his little chin on my arm. As I stroked the faithful little head I wondered and wondered what the future had in store for us. During these long months we had been training steadily while longing to get into action, and now it was coming!

That knowledge brings a grimness to every soldier, no matter whether he be veteran, untried soldier or carefree scallywag. Between the recurring visions of fire and battle my thoughts switched again and again to thoughts of the old folk, of friends, of home and Australia. I just could not sleep those deathly silent hours away.

A match flared up to light a cigarette.

"What's up Don? I murmured. "Can't you sleep either?"

"No," he answered quietly. "I was wondering if that last letter I wrote will arrive home safely."

"Of course it will," I replied.

"It's funny, you know," he murmured, "the little things you try to say in a letter home. You can't. Somehow, they seem sort of sissy on paper."

"That's so."

"Reckon she'll be pretty willing in Greece," ventured Don.

"I think so too."

"Wonder how Poppa will go?"

"He'll be jake; you couldn't kill him with an axe."

"Sorry the Gogg and Big Jim won't be with us at the start. "

"I'm sorry too, but they'll catch up. Nothing could keep those two out of this stunt."

"What about putting on a brew, you blokes?" came Murchie's voice.

"Can't you sleep either?" asked Don.

"No. I can't remember which box I put the Bren magazines in, that I loaded with armour-piercing bullets."

"Still thinking about your old guns."

"No sugar in my tea, thank you!" came from the direction of Fitz's bunk.

"None for me either," called Gordie.

So we got the primus going and soon were sitting around it waiting for the billy to boil. As we were pouring out the tea there came a snorting sniff and Poppa stepped in the tent door.

"Thought I smelled it," he grumbled. "Heard voices too."

"Not mine," called Feathers from his corner. "I was waiting for them to wake me up and offer a cup of tea."

"You lazy leadswinger," said Murchie. "Poling on your mates as usual."

"It's a luxury to get the chance," grinned Feathers. "They're such expert polers themselves."

"Our last night in the land of milk and honey!" droned Poppa sepulchrally.

"Yeh!" drawled Fitz, "if you bring your own bees and cows."

"Oh, don't take any notice of Poppa," said Murchie witheringly. "The only milk he knows is in a beer jug."

"I wasn't dragged up anyway," said Poppa as he squatted down and reached out a horny paw. "Ah! this tea will go down good!"

The great day dawned. A quick, more serious breakfast than usual and the troops stood to their kits in suppressed excitement. Horrie's great moment had also arrived. He was so excited he could not keep still, hurrying from Rebel to Rebel to gaze outside the tent at a thousand men fitting on their packs, handling machine-guns and rifles in queerly realistic fashion. Line after line of loaded trucks rumbled away across the desert to sharp words of command. Troops were now forming up. Horrie waited with eager eyes, alert little body on quaint stubby legs trying hard to stand with martial bearing, proudly dressed up in a new collar flaunting the battalion colour-patch. We kept a sharp eye on him, otherwise he would have dashed out the tent and lined up at the head of the battalion.

"Fall in!" came a shouted order and the Rebels hurriedly grabbed the last of their gear.

"In you go, Horrie!" I ordered, and the dumb look of misery on the little Wog-dog's face hurt me, as quickly he submitted to the indignity of the kit-bag. This was to be no brave show for him; in all his glory he was hidden in the hot, gloomy depths of a kit-bag.

As we filed out for final inspection, it could not be over too soon for me. The ranks stood like statues as the in-specting officers came slowly, ever so slowly, down rank after rank, closely inspecting each man. There could be no replacement of kit from now on; it could be a serious matter if any careless or flurried man left an article of gear behind. The stifling hot day was not the cause of my sweating as those intent officers drew nearer and nearer. Would Horrie bark? Would he growl? Would he shuffle deep down in that stuffy kit-bag? I knew what he was feeling, half suffocated listening to the beloved sounds of the parade around him.

Horrie was a little brick. Not a whimper, not a movement as the inspecting officers ran careful eyes over me and my gear. At last they moved on.

Soon we were on the march across desert to the Ikingi-Maruit railway station.

And fate was kind. A truck came along whose driver was in the know, and I passed the little dog up to him.

"He'll be all right," smiled the driver. "I'm to unload at the station, then goodness knows where I'll be ordered. I'll tie Horrie up to that little tree away back of the station. You know the one I mean."

We did. There was a Rebel story to that same tree.

Horrie was fortunate to dodge that long, hot trek. It would have been awful in the kit-bag.

During the dense muster at the station the Rebels "clocked" me while I quietly edged aside and hurried away to the tree. Horrie was there, a pathetic little figure staring out towards the station. His frantic delight subsided to pleading little whimperings and pawings at my urgent warnings. I bundled him into the kit-bag again and there was not a movement out of him as I hurried back to the station.

Once aboard the train we let him out amongst the packed troops. He dived straight for a window and for the remainder of the trip was the keenest of us all; every-thing in the passing countryside was of interest to him but his delight was to growl and bark at every Arab we passed.

It was a monstrously slow trip. Again and again we stopped to allow the passing of troop-train after troop-train. Something big was doing.

Horrie enjoyed the stops. Urchins and full-grown Arabs would appear at the carriage windows with cries of "Eggs acook!" "Eggs er bread!" hopefully offering us dirty-looking boiled eggs and rolls of alleged bread. That they grinned at Horrie from the safety of the platforms made the Wog-dog all the madder.

With a speculative eye, while disregarding the angry Horrie, a Wog shuffled up to the Rebels' carriage obviously clutching something precious under his dirty rags.

"Good whisky" he whined with a cunning glance around him.

"Shufti, George (show me)!" demanded Murchie and the Wog, casting suspicious looks around shuffled to the window then slipped a bottle of whisky into Murchie's hand. With a frown, Murchie carefully examined the bottle, the cork which certainly had not been tampered with; nor had the label, or the tempting, clear amber fluid within. We were not going to be taken down, no wretched Wog was going to catch us. Silently we awaited Murchie's verdict; this really did look the goods.

"Undoubtedly Scotch!" declared Murchie.

"Hooray!" yelled Poppa, "buy it!"

"How much?" demanded Murchie.

"Five hundred mils (12s.)," replied the Arab.

Murchie paid over, and the Arab with an eye alert for military police disappeared amongst the station crowd.

"Cheap!" declared Murchie, fondly handling the bottle.

"Open her up!" advised Poppa.

Murchie did so with a gusto and offered the first drink to me. I was just in time to intercept and interpret Poppa's urgent wink, so handed the glass back to Murchie with a polite "After you, Dig!"

Murchie slung back his head and took a long, deep gulp that screwed his face into an agonized expression. With bulging cheeks he dived for the window, to Poppa's uproarious mirth.

In surprise we laughed at Murchie's anguished spittings as he howled swear words back to the station as the train moved off. Something obviously was lamentably amiss with that "Undoubted Scotch!"

"Ha, ha, ha!" roared Poppa. "Five hundred mils for a bottle of weak tea - if it's not something a jolly sight worse!"

We looked aghast at the implication while Murchie yelled "Well, it's for, you gloating sewer rat! It's only weak tea!"

We roared, as Murchie glowered at Poppa, who smiled with the knowing expression of the old desert campaigner.

"You don't deserve your luck!" He wagged a finger at Murchie. "It's not often the Wogs fill the bottle with good weak tea. Mostly it's camel's water, or horse, or donkey, or Wog. What I'm surprised at is that dull-witted Wog who picked you didn't sell you a bottle of Wog's!"

"Oh yeh!" spluttered Murchie. "Well now I'll tell you something, you'll be surprised at. That five hundred mils I paid the Wog was the five hundred mils I owed you!"

"You're telling me!" demanded Poppa.

"I'm telling you! Now laugh that off!"

"What I want to know, broke in Feathers, "is how they got the whisky out! I'd swear that that stopper has never been tampered with."

"Neither it has," explained Poppa, "but look at the bottom.

We did so, but could distinguish nothing strange until Poppa pointed out a tiny round mark in the glass on the bottom of the bottle.

"They bore a little hole in the bottom," explained Poppa, "drain the whisky out, then fill the bottle with cold tea or what not. Then they mix up a pinch of borax and soda, heap it around the hole, and melt it into glass with a blow-pipe. The glass bead fills up the hole again. Given a good man at the job, it is practically impossible to detect."

"So that's it!" exclaimed Feathers. "Well, I'm blessed."

"Not quite all of it," said Poppa pityingly. "When the bottle is filled they sell it to some poor mut for five hundred mils."

"Just so," broke in Murchie unforgivingly, "but in this case you are the mut who's paid the mils."

"We'll see about that, me lad!" said Poppa fiercely.

"Oh yeh! How are you going to get it?"

"I'll take it out of your hide!"

"Yeh! You can't give me one more fatigue than I'm willing to do."

"We'll see about that!" promised Poppa darkly.

At long last we reached the wharf at Alexandria, the little Wog-dog a bundle of excitement as I put him back into the kit-bag and petted him warningly. Now was the critical moment. One slip, and the Wog-dog would never see active service with us.

We filed out of the train on to the wharf and formed up in three ranks for the final roll-call. I gazed up at the troopship awaiting us, the crew lining the rails.

On the bows was the dim name Chakla. She was only about three thousand tons. I wondered how on earth we were all going to pack aboard. We thrilled while gazing up at the scars of shell and bomb that had not stopped the little ship. But my anxious concern was to be aboard her. Once up that gangway, now so very close, and the Rebels could vanish in the search for Horrie's "suite". "Once aboard the lugger" kept recurring to my mind.

Sharp and clear rang out the roll-call and man after man smartly answered. No inspection this, so for air I'd allowed Horrie's head out of the bag, his "port-hole" part hidden by my arm. As bad luck would have it just as I opened my mouth to answer my name Horrie spotted an Arab and his bark yelped out with my "Here, sir."

A low chuckle rippled from the ranks, a smothered growl from Horrie. With an innocent expression I cast a glance in the direction of the officers and noted a discreet smile or two. Ah! Horrie in camp had made more friends than we knew. My eyes faced front on the instant and the old heart beat again as the roll-call carried on.

Presently we were filing up the gangway. Only a matter of minutes now. The quarters allotted the signal section were down aft and immediately we clambered down there into the depths I laughed with sheer relief.

"Good lad," smiled Don, "you've done it."

"We've all done it," I replied.

Which was true. I cannot express on paper the willing teamwork of all the Rebels on Horrie's behalf.

The noisy throwing off of our gear, the thumps as men tried bunks and laughed and joked brought a whimper and sympathetic wriggling from the kit-bag.

"We'll let him out a while," I said. "He's safe here; no one here but the boys."

"Let him out," advised Gordie. "No one is likely to look in on us, and if somebody does we can smother him up in a second."

"Give him a breather," they all agreed, "until we have time to scatter and find his suite."

So to Horrie's delight he was free among the litter of gear with his friends, deep down in a little ship's hold.

We'd almost got everything shipshape, every man had found a bunk, the gear was almost all stowed away for the voyage when to a warning hiss I snatched at Horrie and rammed him into the kit-bag; but the little wretch objected and strenuously fought to struggle out again, stern first. A Tommy sailor, unseen away down in the dim end of the hold had been watching proceedings. With uneasy defiance I stared down towards him.

"It's O.K. digger," he grinned. "You can let him out; we've got one too."

8

THE WOG-DOG UNDER FIRE

I SAT on the bunk and laughed. The Rebels joined in. The sailor came forward; it was our first introduction to the typical British Navy man. In the months beginning now, how many, many times we were to bless those same men! Of course they would never have given the dog away; our own sense should have told us that. But, as Poppa so often assured us, we never did have any sense.

The sailor man was unusually interested in Horrie and the Wog-dog took to him at once. It appeared that the mascot they had smuggled aboard was the "dead ring" of Horrie, apparently the same breed. Horrie had a stump for a tail, whereas the ship's dog gloried in a full-length appendage.

"Which the little blighter always carries mast-high when the Jerries tackle us," said the sailor proudly.

That there should be another dog of Horrie's breed surprised us.

"We thought ours was the only one in the world, too," smiled the sailor, "but when you see our Ben you'll know there's another one; ours is as game as a lion and I'll bet yours proves so too. We picked Ben up at Benghazi after the Italians were driven out in the big push of 1941. He'd been abandoned and we couldn't turn the poor little blighter down."

We looked forward to being introduced to Ben. Above all, there was no need to worry over Horrie now, and no need to hunt the ship for a highly problematical "suite".

"You can let him roam about the ship and find his sea-legs," advised the sailor, "as soon as we sail. But keep him below until we do sail, just in case." And with a friendly "We'll be seeing you," he went about his job.

"We're set," declared Feathers. "The crew are all good blokes. This is going to be a happy ship."

And it was.

When well out at sea next morning we brought Horrie up on deck. He took an intense interest in everything; he'd never been aboard before and exploring the ship was great fun but the waters rushing past had him tricked. Cautiously he approached the rails, always ready to leap back; he could not make it out at all, so he back-pedalled up the deck again. Everyone received a sniff and wag of approval and an invitation for a romp around the deck. When the crew produced Ben the packed deck enjoyed a pantomime while the two made friends.

Horrie's envious interest centred on Ben's long tail which Ben erected mast-high for inspection. Ben reciprocated by polite but amused interest in Horrie's stub of a tail and they slowly waddled around one another in a mutual admiration society. Then Ben invited "Come along; I'll show you around the ship!" and waddled off with Horrie staggering at his tail. The little ship was now rolling uncomfortably and there were unhappy signs that many of the soldiers shared Horrie's feelings.

But Horrie had not yet ceased to take an interest in life. When the dogs returned to the deck Horrie was proudly struggling with a bone as large as himself. On their tour of inspection, the first place Ben had introduced him to was the ship's galley and the cook. Just as well, for we only had bully beef to offer him.

As a reward for the bone, Ben expected Horrie to romp and was obviously contemptuous of his efforts. Horrie tried hard but vainly, as he now rolled and staggered with a surprised uneasy expression on his face. He could not understand why his legs were giving way under him, could not make out the sickly feeling stealing over him.

After midday it blew up rough and Horrie lost all interest in fun and games. I never before realized the truth of the expression "sick as a dog". I selected a quiet place to lie down in and die, and Horrie crawled up to me; he could not walk. Surely there never was a more miserable pair. Horrie with his chin rolled on the deck, his little legs stretched out fore and aft, his eyes gazing for hours into my face, beseeching me to do something to stop the ship rolling. I would have moved heaven and earth to have been able to do it, but all I moved was my dinner, which hurriedly went to the fishes.

The crew were wonderful. To anyone really ill they immediately gave up their own bunks, while cheery words and a hot cup of tea were readily available.

Next morning was much calmer, thank heaven. Ben came waddling along to coax Horrie to come and play but he miserably declined. He lay there with envious eyes silently watching the pranks of the old sea-dog.

A sailor asked me, "Can I show your Horrie to the Old Man?"

"Oh, Poppa's seen him often," answered Murchie.

"You take the Wog-dog to the Old Man," said Poppa to the puzzled sailor, "take no notice of this has-been."

"Oh yeh!" growled Murchie.

And Horrie waddled away to be introduced to the Captain while Poppa quietly went below. He soon returned and sitting down beside the helpless Murchie began with gusto to suck a large piece of greasy fat. A ghastly sound.

Murchie rolled towards me. "Chuck this --- over-board!" he implored weakly.

Poppa laughed and gurgled.

"What! Not seasick, surely, Murchie?"

"I wish the ship would sink and you too - especially you," moaned Murchie.

"She won't sink," said Poppa confidently. "Don't worry, she's a good ship." And he sucked with a sound that --

"Please go away!" I whispered.

"If only she would sink," groaned Murchie.

"She won't sink," declared Poppa brightly. "Don't worry, Murchie boy."

"That's what I'm worrying about," groaned Murchie. She won't sink.

But the rolling calmed down and gradually we revived. In late afternoon we passed several islands at which Horrie, rapidly recovering, gazed with earnest attention.

"How he'd like to be ashore," smiled Fitz.

"No more than me," said Murchie feelingly.

"We're getting near Bomb Alley, lads!" called a sailor warningly.

Neither Gordie nor Fitz nor Murchie nor Feathers nor I were interested in Bomb Alley or any other alley. But we presently realized that soldiers who had not obeyed the warning to wear life-jackets at all times were now attempting to put them on. The Chakla had made numerous crossings between Alexandria and Greece carrying Australian soldiers and had been heavily bombed in some place called Bomb Alley.

"Better put your life-jacket on, Fitz," I suggested.

"Give mine to Horrie," replied Fitz. This jolt brought us back to life; we'd forgotten all about a life-jacket for Horrie should the ship be sunk.

"Could a pup swim in a sea like this?" I asked Poppa.

"Not for long," he replied doubtfully.

"Couldn't we fit Horrie up with a life-jacket?" suggested Don.

I remembered a broken, discarded jacket and wandered away to get it. We cut three pieces of cork from it and sewed them into a strip of canvas making the jacket into the form of a saddle, one strip to fit down each side and one over the back. I called Horrie, who was playing with Ben. Horrie came inquiringly across the deck but did not seem pleased when we fitted the belt around him. Ben stood with a wary eye on proceedings. Horrie insisted on wagging his tail at Ben which made the fitting difficult. Don hacked a half-moon out of the centre piece of cork and this enabled the jacket to be pulled well forward, the idea being to keep his head more buoyant than his stem.

"What he swallows in the stern won't matter," said Feathers.

The final problem was to fasten the jacket in place. "String would cut into his tummy," objected Gordie.

"Half a mo!" advised a sailor. He returned with a smile and a long bandage - the very thing. We wound the bandage over the jacket and round and round Horrie's middle, then fastened it securely in place. A perfect job.

"It's the first dog's life-jacket ever made," said Murchie admiringly.

"We ought to go into business," declared Gordie.

We'd almost forgotten our seasickness.

"Off you go!" I ordered, and Horrie scampered away to his waiting friend but feeling the life jacket rolled over and energetically kicked out. This effort proving vain, he leapt up and did a buckjumping turn at which Ben barked approval, encouraged by sickly laughs from the soldiers. Horrie when out of breath invited Ben to give a hand and Ben promptly fastened his teeth to the bandage and bracing his legs tugged with might and main and brought Horrie down on top of him. To a startled yelp and struggle from Horrie they sorted themselves out. Then Ben noticed the end of the bandage had become unfastened so he snapped at it and tugged again while I urgently called Horrie. Horrie struggled to obey but was pulled back by Ben. Horrie indignantly wheeled around and snapped at Ben as if blaming him for the laughs at his expense. Then he came trotting across to me but forgave Ben and turned to bark that he "wouldn't be long now". As he turned back towards me Ben grabbed the bandage end and bolted down the deck. And that was the end of the safety jacket.

That evening we steamed into a pretty harbour with numerous small islands and buoys dotted about and - what kindled the troops' enthusiasm after so many months in the desert - beautiful green foliage lining the shore. Bracken fern and green grass and clean white houses, made a lovely sight after the desert. This was Piraeus. If ever troops welcomed a port, we welcomed the Port of Athens.

The Wog-dog who was barking his joy to the shore, was hurriedly popped inside the kit-bag. Ben thought it a new game and playfully nipped the bulging bag, to an agonized howl from Horrie. This was a bad start as we were not lining up for disembarkation, but Horrie, stung to the quick by the unfair attack, was determined to bark his way off the ship. I was trying to stand at ease while the gangway was lowered but Ben waddled around and barked up at Horrie's muffled barks. The crew enjoyed themselves but I was sweating, for the chorus of barks was creating more attention than the orders of the commanding officer now giving us disembarkation orders without, it would seem, even dreaming there was a Wog-dog on board. Thankfully I saw a fleet of small lighters drawing alongside the ship. We began to clamber cautiously down the gangways.

From the lower end of the gangway to the small craft bobbing on the swell below was a precarious drop of three feet or more. A man had to pick his time and, helped by the outstretched arms of the Greek sailors, to jump. If he missed the jump he went into the Drink.

The cheery Greeks were helping the boys by catching the kit-bags dropped to them. I had no option so carefully dropped my bag into the outstretched arms of a Greek sailor and then jumped. As the Greek clutched the bag Horrie wriggled and growled and snapped and the surprised Greek toppled backward. I snatched the kit-bag and bustled in among the boys. The sailor picked himself up with a surprised expression that changed to alarm as the air-raid siren above us whistled piercingly. To urgent shouts and gesticulation among the Greeks our lighter steered an erratic course towards shore, sped by the bomb bursts and shrieks of diving planes. When we leapt ashore Horrie sprang out of the bag with a bark of delight and led the scramble for cover. The Greek anti-aircraft guns crashed into action and Horrie leaped around as if thunderstruck and scampered to me terrified. I snatched him up and raced to a drain into which the boys were disappearing. Horrie was trembling like a leaf. We crouched in the mouth of the drain and watched the raid while I tried to soothe Horrie. Ships were fighting for their lives, searchlights swept the sky to catch a gleaming plane that missed, turned and dived while shooting machine-gun bullets down the beam of light. As plane after plane was caught in the searchlights the sky broke into a roar of bursting shells intermingled with the red and yellow and green of flying onions and streams of tracer bullets, a terrifyingly beautiful spectacle. Then came the deep "whoomph!" of a bomb landing fairly close and Horrie with bristling hair leapt into action to bark furiously at the silver planes in the sky. With tail erect and eyes fiery gold he barked his defiance to the Luftwaffe away up there dodging the slowly moving streams of fiery onions that were now bursting from every point of the compass. Horrie had made good.

An hour later and we trailed off through the night to the little Greek village of Dophney. During the early hours, low-flying, hostile aircraft dropped their eggs with a roar that shook the ground, showering leaves on the tent roof. Dazed by the fiery explosions the little Wog-dog crept to my bunk and buried his cold nose under my armpit. Presently he gained courage again and growled, a low growl, but I'm afraid he did not really mean it.

"Just you come down here so as Horrie can get his teeth into you!" yelled Murchie as an enemy plane roared overhead.

"He can't put the wind up Horrie," came Feathers's voice from the darkness, "like he's got it up me."

"Hah!" came Poppa's growl. "Soldiers!"

"The old relic is keeping his courage up," came Fitz's distant voice. I yelled out for Gordie and caught a muffled answer. The Rebels were all right so far.

Dawn, smiling over sea and hills, found us in a delightful spot. Green fields, green crops, well-tended vineyards, fields of poppies, little white houses all welcoming a glorious April sunshine. Then Feathers called out in delight.

"Good old gums! Well I'm blessed! We're camped under Aussie gum-trees."

And so we were - a good omen, the familiar, dear old trees to shelter and welcome us.

"This is a country worth fighting for," declared Gordie. And we all agreed, gazing over the lovely scene. And so this was Greece.

Lots of smiling, dark-eyed youngsters soon came to invade the camp. We made good friends. Who can help making friends with youngsters? No man but could help be bashfully flattered when these youngsters looked upon us as heroes come to save their land.

Alas.

A particular delight to the kiddies were the tins of bully beef we gave them. Meat of any kind seemed very scarce, a luxury of luxuries to the youngsters. Here too the Wog-dog had his first offering of bully-beef and - turned up his nose.

Next morning we passed in triumphal procession through cheering Athens. The people went almost frantic; they rained flowers upon us from streets and balconies. The little army of these folk, so very, very few, had held back the might of Italy for many long, bloody months. And now just as the hordes of Germany were marching to overwhelm them, here were we come to save them. As our convoys of trucks rolled through the streets the soldiers scrambled to the truck roofs, swarmed on the mudguards, catching the flowers thrown by the people, giving cheer for cheer in tumultuous greeting. On the Rebels' truck Horrie was wildly excited, barking from the roof of the truck at the footpaths and up to the balconies. In the roars of cheering and laughter and whistling he was like a cat on hot bricks not knowing which way to turn. The sun shone smiling upon this lovely land, and these warm-hearted people. Lots and lots of folk laughed and waved to the little excited pup. Their wave of greeting is unlike our own. They extended the hand with the palm turned up, closing, opening and closing the hand more in a beckoning attitude than a wave. The meaning was similar to our "Good luck", but meant more precisely "Go and return!"

"I wish we were the last truck in the convoy!" sighed Don.

And Feathers in spick and span uniform had smiling eyes all for the laughing dark-eyed girls crowing around the slowly moving trucks.

"I'd shed no tears," said Gordie thoughtfully, "if we were left behind amongst this crowd."

"A flat tyre would be welcome right now," mused Fitz.

Murchie was balancing precariously on the truck roof enthusiastically waving his rifle to the shouts of "Benghazi! Benghazi! Benghazi!" The fall of Benghazi to the Australian troops was a beacon light in the minds of the Greek people, and here was Murchie taking all the credit.

Alas!

Our reception was that of conquering heroes. Perhaps the Greeks knew better what lay ahead of us, realized more than we. Surely they had hoped for a far greater force. The story of David and Goliath is not repeated in modern warfare when it is a handful against many men and many machines. We were to give our very best but --

At last we were clear of the city.

We gazed back at Athens. High up on a hill the ruins of the famous Acropolis seemed brooding over the city. Then we were travelling along the northern road, going inland, winding through village after village with quaint square-built white-washed houses, each little village with its own church with high steeple, each with its little village square with a white wooden cross in the centre. The trucks rumbled along cobbled streets, sometimes so narrow there was only room for a military truck to pass. The footpaths were crowded with cheering villagers, the windows in the dwellings above the shops all with their laughing people showering flowers and good wishes down upon us. Most of the folk wore their national costumes, and the village lasses were a picture. Jet-black hair and laughing brown eyes were emphasized by the smooth, olive skin. They were dressed in white, ankle-length skirts, embroidered with flowers of contrasting colours - lively young things in their white lacy blouses pulled in tight around tempting waists and little vests of vivid blue or red laced crisscross down the front with silk cord. Strong, shapely arms showed from wide, puffy sleeves caught in tight above the elbows. Little sandalled feet peeped from below generous skirts. The men folk were mostly at the front, but those that remained wore long sheepskin coats and black velvet trousers caught in below the knee after the style of "bowyangs", their headgear small skull caps. Here and there was a Greek priest in long, black gown. Nearly all the men wore short beards.

Horrie was almost exhausted with excitement. As we passed through village after village the funny little dog frantically barking and dancing up on the roof of the truck was a source of attraction and laughter.

"He thinks the whole show is being put on for him," laughed Don.

"If only I could command as much attention!" sighed Murchie as we saluted a gay bevy of beauty.

"You're attracting plenty of attention," growled Poppa, "they haven't seen such an ugly man since Caesar passed through."

But Murchie was too busy with the girls to reply.

9

THE WOG-DOG
CONQUERS GREECE

EVER four miles along the road we would pass a white-washed box set upon a post. Through the glass front a little lamp would be steadily burning at the foot of a miniature crucifix fronting a painting of the travellers' patron saint - St Francis, I believe. A till in each box contained a few coins. The shrine offered any traveller the opportunity of worship and, if needed, the money to buy food along the road. Peaceful symbols, these, in a lovely country soon to be trampled under the iron-shod heel of the invader.

As we passed through Larissa Horrie barked frantically at the sky and, leaping up, overbalanced to tumble into the outspread arms of folk below.

"What bit him then, I wonder," said Fitz as we glanced up to see a stork perched on its huge nest high up on a church steeple. The solemn stork gazing so calmly down on the laughing people, the excited little dog made a scene that lingers sweetly in memory.

That night we camped off the road by the side of a rocky hill, green under bramble, gay with red poppies and white flowers. Through a valley a stream flowed over a bed of snowy pebbles. Along the banks on either side little fields of green and golden barley rippled under a whispering breeze.

"It is a beautiful country," said Don that evening.

The Rebels were unusually silent, gazing up at the golden stars, dreamily aware of the tinkle of tiny bells on the necks of sheep and goats wandering upon the hills.

"Yes, it's a lovely country," sighed Poppa and we felt sadness in his voice.

"It's not the first time that armies have marched through here," protested Gordie, "and the country and people still survive."

"They'll survive this one too!" declared Feathers.

We were silent, only just beginning to realize what our small force was to face.

Next day Don and I were ordered to travel by cycle, so Horrie was parked inside my greatcoat which I belted so that he could not fall through. He insisted upon seeing all that was to be seen and poked his little head out between two buttons, to the delight of the Greek folk whose attention he attracted by his ever-ready bark. He was to travel very many miles in this fashion, some of them awful miles, the terrible miles of a retreating army ploughing back through hail and mud under a pitiless rain of bombs and shells. I had only to sit on the cycle and unbutton the coat and Horrie would scramble up and dive in like a baby kangaroo into the pouch. In a moment his little head would be poking out and he was ready for the road. The instant I stopped, however, he would scramble out.

That first morning we travelled on and up through the Thermopylae Pass of deathless memory. The road wound round and round as we climbed. At the top we gazed down upon a beautiful scene. Nine sections of the road were clearly visible, each at a different height as it climbed the hills. Far below the green fields were spread out like a chessboard with little white doll's-houses of farms. A long, perfect stretch of road gleamed like silver in the light as it cleanly divided the chessboard from the bottom of the Pass. Towering distantly above this scene gleamed the snowy crown of Mount Olympus.

The day grew very cold. I felt the Wog-dog draw deep down in the coat until only the tip of his nose was visible. Then that, too, disappeared.

"I feel Horrie shivering," I said.

"I'm a bit that way myself," replied Don, "but we must fix Horrie up. I've an idea."

"What is it?"

"I've got a couple of spare socks. We'll cut the heels and toes off and draw the remainder over his body."

"He'll be snug as a bug in a rug," I laughed. "Let's do it.

So we dismounted and very soon fitted Horrie with his "nightie". It was a tight fit; he looked comically uncom-fortable but as he felt the warmth his tail began to wag and he gazed up with appreciative brown eyes, plainly saying "Thanks boys. you two certainly do know just what a fellow needs."

We were surprised that no aircraft had yet had a shot at us but were to discover that torrential rains had hopelessly bogged the enemy air force. That rain saved us many lives.

Time found us establishing battalion headquarters at the foot of Mount Olympus towering nearly ten thousand feet above, capped with snow that gleamed through the purple haze. Around us rose the majestic mountains of Greece. Through rock-walled passes ahead came the echoing thunder of guns. Around us thousands of men were busy with a grim earnestness. Along the rocky road a stream of trucks was arriving, loaded with materials for the gluttonous maw of war.

"We're going to be busy soon!" observed Fitz.

"You're busy now!" called Poppa. "Come and give a hand with this gear."

"The old warhorse smells action," nodded Gordie.

"We all will soon," said Murchie eagerly, "judging by that nasty noise coming over the mountains."

"We'll be in a dangerous position here," remarked an artilleryman, "if the Greeks collapse away out on the left flank."

"Poor beggars," said a weary mud-covered soldier. "They've fought like Trojans throughout months of it and now must stand up against the German panzers and Stukas."

Sergeant Poppa, working like a Trojan and supervising signal communication, came hurrying across with orders.

"Fitz and Gordie to remain temporarily at B.H.Q. for signal duties," he ordered. "Feathers to go with A Company in charge of a section of attached signallers. Murchie to be spotter on the adjutant's car. Don and Moody for dispatch-riding duty under Corporal Thurgood."

"And what are you going to do?" inquired Murchie.

"I am taking over the duties of signal officer and signal sergeant as well!" snapped Poppa. "Got anything against it?"

"Oh, no," drawled Murchie as he lit a cigarette. "Neither will the enemy."

"You won't be feeling quite so smart a few hours from now, my bright lad," promised Poppa grimly and hurried away to his duties followed by our grins. A great old scout, Sergeant Poppa.

Presently, D Company moved out towards the fighting at the Albanian border while A, B and C Companies moved out to join two Australian and a New Zealand Brigade now in heavy action defending Servia Pass, some sixteen miles further ahead. We stood silently and watched our mates hurrying towards the hoarse rumbling of the guns.

Next morning early I received orders to rush a message to the firing-line. I felt a funny little tremor away down in the pit of my stomach.

The mud-caked, grim-faced wounded spoke without words of what was happening away up in the iron-jawed heads of the pass.

The Rebels were all away at various jobs; any of the lads would gladly have kept an eye on Horrie but I determined to take him with me; I couldn't have faced the Rebels had anything happened to him while he was in my charge.

"We're in for it now, Horrie!" I said, and patted the little Wog-dog.

We mounted, and started up the long, historic road. Horrie, peeping from my greatcoat was this day to see more sights, hear more things than even his insatiable interest and curiosity could stomach. The road proved muddy as we rode on to the rumble of the guns. Burned-out trucks and smashed flotsam of war littered the roadside. Progress became slower and slower. I was riding cautiously, staring at the mouth of the pass looming in front, wondering why the stretch of road just ahead was ominously deserted. The noise of the cycle's engine muffled sound until a screaming roar lifted the hair on my scalp. The mud spat up to machine-gun bullets as I hauled the cycle over. We crashed into the ditch as the diving plane roared past. He zoomed up but another was roaring down as my finger felt for the little Wog-dog's head.

"Are you all right, Horrie?" I whispered. Breathlessly I squeezed into the mud as the second plane screeched down with machine-gun bullets hissing around us. Horrie kicked frantically under me as the third plane roared down and for a second I thought he was hit. But it was the spill into the ditch and the scream of the planes that had terrified him. They had scared me, let alone the little dog.

The three planes as they zoomed up circled lazily over-head. I peered up anxiously and saw the devils' heads gazing down from the machines; I could visualize the grins on the cynical, masked faces. The machines began to climb in a horribly suggestive circle. I scrambled up and seized the bike to race for cover before they could dive again. Horrie whimpered in sympathetic understanding that something tragic was doing. My heart thumped violently at a screech of brakes then a voice called down from a big truck.

"Are you all right, Dig?"

"Yes, no, yes, no," I grinned, "you nearly frightened the life out of me. I'm a bit winded, that's all. Those planes spilled me into the ditch."

"It's lucky the ditch was handy," he said. "Sure you're all right?"

"I think so," I replied hopefully.

"Then race for the pass," he advised. "They'll be back in a minute." And the big truck lumbered off.

I leaped on the bike and was away with an eye on the three planes climbing and turning away to the left as the pilots gazed down in search of victims. There was a half-mile of straight road between me and the pass. I leaned over the handle bars and went flat out, heartened by the roar of truck engines as a convoy appeared behind me. Spotters clung to the running boards, searching the skies.

"Still, Horrie; keep still!" I implored. "Good dog, good dog!" Horrie was very nervous and should he leap out I didn't know what might happen. The road was awful - pitted with bomb holes, greasy with mud. We couldn't make it but this time I heard them and slewing the bike off the road jumped off and tumbled into a ditch, hugging Horrie. "Thank heaven!" I muttered. "We beat them that time!"

The trucks had pulled up with a screech of brakes while all hands jumped down and raced to cover. In an instant the road was devoid of life. There was just a long line of stationary trucks. Above them were the diving planes with sirens screaming, machine-guns "rat-tat-tatting", answered now by the vicious crackle of rifle-fire from the men lying in the fields.

And so it went on. The enemy planes were enjoying the game; they'd zoom up and circle as if to fly away. Then when the trucks got going again they'd suddenly wheel and dive straight down.

With each run off and dive into the ditch, Horrie began to regain confidence. His ears pricked up and I could feel his stub of a tail wagging as we dived off the road and he leapt out to cover with excited barks.

Apparently the planes were short of bombs though once they dropped two some distance away. Horrie and I were racing and fell into a furrow and stayed there feeling big as pyramids while waiting for the worst. The play went out of Horrie to the shattering roar of the explosions. Pressed into the furrow his eyes glared up at the planes and his hair rose on end. He growled as ferociously as a very small dog can possibly growl. Those bombs obviously took his memory back to the raid on Piraeus; he knew now this was no game. When silence reigned I peeped up and laughed to see heads popping up all over the ploughed field, all eyes watching the receding planes and almost everyone saying the same thing:

"Missed, you b---s, missed!"

I held up Horrie to gaze after the vanishing planes; his eyes fairly spat fire and he growled menacingly to the relieved laughter from the boys.

From that moment Horrie became the troops' best spotter, wherever he might be. He was immediately to become famous at detecting approaching enemy planes. In all our various movements, throughout all the fighting that we saw after this Horrie would always be the first to pick up sounds of approaching aeroplanes and some among us would always keep an eye upon Horrie. "Planes!" a man would shout and we'd dive for cover. For Horrie would have been noticed sitting very erect with his ears cocked, a look of intense interest on his face. Immediately that he growled then the cry would arise, "Planes!"

But this was Horrie's first morning in battle apart from the raid on Piraeus. At least, the three planes did fly away and as we were breathing sighs of relief their place was suddenly taken by twenty-one devils that dived down with the scream of bombs howling above the screech of sirens.

I called to Horrie as he led the way and we dived into a hole as a bomb exploded with a shattering roar. The Wog-dog pressed himself against me, his little heart thumping.

"Good dog," I said encouragingly, "keep low; it will soon be over."

Bomb after bomb shattered the crackle of rifle-fire as the machines roared low over the convoy, spraying it with lead. As they sped away for more bombs men leapt from the earth and ran for trucks and within seconds the convoy was heading for the mouth of the pass, now very close.

But the frightful noise and grim fear of death had again put the wind up the Wog-dog. As for me - - !

We gained the mouth of the pass. The gloomy iron walls of the mountains now sheltered us in some degree from the planes. The road now was a nightmare, pock-marked with shell and bomb craters, the mud from which was a quagmire. The sight of the once pretty little village of Servia clinging on the western side of Mount Olympus would turn any man against war; it was a rubble of smoking ruins. Families of warm-hearted people had lived and loved and toiled in that village only a matter of days before.

As I passed the ruined village, I was halted by one of the road patrol.

"Be careful, Dig," he advised. "Jerry is shelling the pass again. Been through before?"

"No."

"Well, here's the drum (good oil). Work your way quickly close up beside that second bend, see, away ahead!"

"Yes."

"Well, when you get there keep close to cover and watch the head of the pass. Stay there until you see the flashes of the gun, then beat it through the bend before they fire the second shell. Don't worry about the first shell, it'll be at you soon as you see the flash. If it misses, then go for your life."

"Thanks," I said.

"Where did you get the pup?" he grinned.

"At the first flash?" I replied from a single track mind.

"Not scared, are you?" he asked a little anxiously.

"No," I lied. "Oh, the pup! We brought him from the desert outside Alexandria."

"You don't say! How on earth did you manage to get him here?"

"Oh, we got help all along the line; lots of the boys helped us out of little difficulties."

"I suppose they would," he said. "He's a bonzer little pup. He'll make a good little soldier too!"

"He'll be O.K." I replied. 'Thanks, Dig," and I rode -on, leaving him to his dangerous post. He seemed to be an accusing sort of chap, but I'll admit I did have the wind up a bit. Well, just a little bit.

I reached the second bend safely and pulled up along-side a stationary truck. The driver was standing by the side of the truck, dreamily gazing up towards the head of the pass.

"How's she going?" I asked.

"Not bad, Dig," he replied absent mindedly. "The first one's coming!" He nodded as we both dived into a ditch hurried by a heavy explosion over the embankment. "This one should be in the I diddle diddle (middle of the ditch)!" he said, as he wiped mud out of his eye.

Thank heavens his judgment was a few yards out; the shell landed with a crash just over the embankment a few yards to our left.

10

THE WOG-DOG SAVES LIVES

"WE'RE jake (all right) now for a bit," he drawled and lazily stood up, "although you can't bank on it - Jerry is using a pretty big mortar and occasionally whips in a quick - fourth for good measure."

"That's the one that might catch a man."

"Yes," he grinned, "Jerry knows all the tricks."

I crawled out from the ditch holding my hands reassuringly over the greatcoat.

"Hullo!" exclaimed my muddy friend, "you look like you're in the family way."

"It's my little dog," I grinned and took Horrie out from the coat. His tongue tried hard to lick me, his tail still had a waggle. He had a wag too for the driver who grinned and patted him.

"Look, Dig," he suggested, "the road is knocked about pretty badly around the next bend; you'll be kept very busy managing that machine and dodging Jerry's shellfire; you'll get it hot and strong soon as we poke our nose round the bend. Better let me take the pup in the truck and I'll pick you up at the end of the pass."

Thankfully I accepted the offer and Horrie, to his dismay, vanished into the truck which started immediately. As I pushed the bike to the road I saw Horrie's little face anxiously pressed to the window to see if I was coming. I leapt on and waved to him; then all my time was taken up watching the greasy, broken road, leaning a little more closely over the handle bars. Again there came that "Whiz - zzz!" with a crash to the side or behind me. There was no time to dodge the shells; it was just a race to get through the pass. I'll never forget the feeling of relief when we managed it and got shelter from that merciless confined shelling. From close by in the mountains came the sustained roar of rifle and machine-gun fire but it did not have the terrors of the whining howl of shell splinters, deep down in Hellfire Pass.

Horrie did not wait to be let out of the truck; he jumped from the window.

"That little bloke was very worried about you," laughed the driver.

"Yes," I said as Horrie nearly tried to eat me.

"Well, good luck, Dig," smiled the driver as he bent to his controls.

"Thanks a lot."

"She's right," he nodded, as the truck moved away.

Out among the hills I located B Company, reported to the sigs, and carried on with my job.

"You got the pup through," grinned Nippy Burke.

"Yes, by a stroke of luck. But I wish I could have left him with the Rebels; I'm afraid I mightn't get him back again. Goodness knows what jobs I'll be detailed to out here."

"I'll take him back to the boys, if you like, when I return," volunteered Ron Baker. Ron was our signal truck-driver and I accepted his offer gladly. The little Wog-dog had endeared himself to all of us; I hated to think of him mangled in the mud among these grim black mountains. Back at B.H.Q. actually was no safer, but there at least he would be among the boys; at least some or other of the boys would be always on hand.

"You've been experiencing a torrid time," I said.

"Yes," grinned Bill Arrowsmith, "we've had our moments," and he ruefully felt a busted nose.

"What gave you that lump on the head?" I asked Nippy. "You look as if you'd been hit by a shell."

"It felt worse than that," grinned Nippy. "The planes came very suddenly and strafed us. I jumped for that big olive-tree and dived around it fair on to Bill's head as he scooted round the other side."

"It was the funniest thing out," laughed Ron. "I saw the crash from another tree and was laughing fit to bust when I felt a little tug at my tail. I stopped laughing when I saw what it was!" and he showed me a bullet hole drilled low down through the tail of his coat.

"A near miss, that," grinned Bill.

"It was," said Ron thoughtfully. "I hate to think what would have happened if I'd been bending."

We carried on with our various jobs while ahead of us slowly grew stronger and stronger the grim pressure of war. Far above, through purple haze there now and then gleamed the snow cap of Mount Olympus overlooking this passing battle, as it had overlooked so very many battles that throughout history had surged down the Pass of Servia. What terrible tragedies have taken place there! How many times has a forlorn little army battled to hold that pass against overwhelming odds, just as our army was bleeding to hold it today? How thankful we felt that our nation, though small in number, lives in a great continent with mighty seas around it. But the unfortunate Greeks dwell in a tiny area of land pressed in by hungry nations of great power. We in our happy country have not the faintest idea of what unhappy little nations like the Greeks have suffered throughout the centuries.

Ron Baker set out on the return trip with Horrie, the little Wog-dog very anxious as he gazed back imploringly at me. This hurry-scurry of war, these breath-taking movements, these waves of terrifying sound with the grim atmosphere of tragedy and tension were very different to the martial tramp of the carefree parades.

The little Wog-dog sensed this was a phase of soldiering he must try very hard to understand.

To my great relief I was soon afterwards ordered to carry a message back to B.H.Q. Soon the bike and I were racing down the pass again; it seemed easier now I had been through it once, but I raced on trying to overtake the truck. When through the pass, I saw the truck in the distance, tearing across the straight stretch beyond. Suddenly it pulled up and Ron scrambled out of the truck with the Wog-dog under his arm. They dived to shelter as the planes roared down. I dived too. Three times again those wretched planes forced us to dive for cover. Finally I overtook the truck just as it entered the bushy track leading to B.H.Q. Ron was just in time to grab Horrie as he leapt at the cabin window when he saw me.

"He's nearly twisted his neck trying to gaze back along the road, laughed Ron.

With Horrie trotting ahead, tail mast-high, we visited the cook's quarters and forgot the terrors of the pass in ravenous enjoyment of a meal.

That night Murchie strolled in with his cheery old grin and swapped experiences of the pass.

"So you've been through it too!" he said to Horrie. "You're a full blown soldier now."

Poppa came in soon after and was joyfully greeted by Horrie.

"It's all right, Horrie," said Poppa, "we're all going to answer roll-call this time; the rest of the boys will blow in soon. Except Fitz and Gordie," he explained to us. "They've gone out for the time being to replace Harry Doran and Mat Taylor. The poor lads have been hit."

"Rotten luck," said Murchie quietly. "God blokes too."

Don and I rigged a little shelter from the bitter night with our gas capes and water-proof groundsheets. There was just sufficient room to crawl in under the cover. You can bet Horrie crawled in too and cuddled up at our feet.

That night we quietly decided that things obviously were very serious. There was no telling what would happen. If there was any confusion and the Rebels were split up, with none of us able to look after Horrie, we would give him to Doc. Sholto Douglas to look after. We knew he would keep an eye on Horrie although he was now very busy. But he was one of Horrie's many friends and the medical unit would always be kept intact, or as intact as the fortunes of war would allow.

We would thus know where to find Horrie again, should the worst happen.

We did not care even to voice our thoughts that it was possible the very worst might happen.

Morning found our shelter crisp with icicles, cold diamonds in the brilliantly bitter morning.

"I hate to think of wounded men lying out in this," murmured Don.

"They'd be asleep long ago," I replied.

We crept out of shelter while the tip of Horrie's nose peeped out of the blankets to watch us. But he didn't move. Not he; he stayed there snug and hidden under the warm blankets for another two hours.

Poppa was away bright and early on his cycle; Don watched him disappearing down the track leading to the pass.

"He'll be all right," said Murchie. "Couldn't kill him with a ton of bricks. The old war-horse will see us all out."

Don laughed. The Rebels would take it hard should anything happen to Sergeant Poppa.

A few hours later we saw him returning with a passenger but he helped him to the R.A.P. tent.

"Who did you bring in?" I inquired as he wheeled his bike back towards us.

"Yugoslav," he replied - somewhat defiantly I thought. "Picked him up this side of the pass. Leg pretty badly mangled."

Later we found out that Poppa's "Yugoslav" wore an Iron Cross on the left breast of his tunic. But Poppa only grinned. Doc. Douglas treated the German and he was sent to the rear in an ambulance. Being detailed for a job I walked to the bushes where my cycle was concealed. Horrie struggled in Don's arms.

"He's ready for another job," laughed Don. "He's been through his baptism of fire and wants another go."

"He'll get more than he wants here," I replied as I mounted. "Don't let him follow me. I don't want to take him through the pass again. It's getting stickier every hour."

The Wog-dog blossomed into a war-dog as if by magic, despite the fact that we long since had ceased to wonder at his instinct and intelligence and adaptability. He knew the whistle of bullets, the whine of shells, the scream of a bomb, he knew exactly what to do. In a moment he could act at some sudden alarm; he knew at any time what was doing and was intelligently ready to take his place at the right time and in the right way. The Wog-dog grew very dear to all of us.

But he saved us trouble and casualties many a time. Instinctively, although each going quietly about our jobs on what appeared a most peaceful morning, always a man or two would have his eye on the Wog-dog. Immediately that questioning little head went up, immediately he sat back and gazed steadfastly into the skies some man or other would stand and stare at Horrie. Should the dog bark then instantly the shout "Planes!" arose and Horrie would be leading the stampede for the nearest slit trench to jump in and bark a warning for us to follow quickly. After the raid was over Horrie would receive and graciously accept numerous pats for his timely warning. Very soon we all learned not to laugh at him, for despite everyone's danger the impulse to laugh was almost irresistible. He was such a comical little fellow; his dignity and his every movement were comical especially when racing to lead us all to safety.

We'd long since known the little dog was deeply sensitive but never dreamed how sensitive until we were crossing the log. This was over an icy mountain stream. A tree that had fallen across was our only method of crossing, but that tree was very slippery. The Wog-dog of course took the lead to show us how to cross. He trotted to the log with tail mast-high, dignity on his funny little face. When half-way across the log he slipped and splashed down into the stream amid roars of laughter. Swimming to the opposite bank he scrambled out, shook himself, then in very dignified fashion stepped on to the log and advanced towards us. When half-way across he stopped, gazed at us, and barked. When satisfied that all were watching him he deliberately jumped into the stream and swam ashore. He made his act perfectly plain; he had not fallen in at all, he did it on purpose to make us laugh.

After that we never laughed at him unless he was obviously playing. We thereafter respected Horrie's dignity, - which was very precious to him.

The tide of war rolled on. The enemy had brought up great weight of reinforcements and metal, determined to overwhelm us, roll down the pass and flood out upon Greece. Our reinforcements were pitifully small while as to our metal and air cover--!

The Anzac Corps felt it was to be a fight to the death when rumours came fast that our Greek Allies, bravely struggling away on our left flank, were in bad shape; their line had been breached in a number of places; if we did not get reinforcements and guns and munitions to them quickly there was danger of them collapsing. There were no reinforcements, nor guns, nor munitions.

The thunder of gunfire at the head of the pass had grown into an increasingly threatening roar.

At times, the sky seemed full of enemy planes shrieking down upon us.

Wounded, utterly weary, exhausted Greek soldiers in ones and twos, then threes and fours, began to stagger in towards us from the Albanian flank.

The Beginning of the End.

11

THE HELL OF WAR

WHEN the withdrawal commenced it fell to the Rebels among others to keep the trucks moving. Swarms of trucks came down from the mountains and the Albanian front to converge in the pass, trucks loaded with wounded men, with ammunition and gear and stores, with all the flotsam of an army in stubborn, fighting retreat. We rode up and down the long convoy, sorting out tangles and bottlenecks, guiding the harassed drivers, directing them at turn-offs. Ray Thurgood worked like a tiger, setting us an example we will never forget. Other men were toiling similarly far up and down the struggling convoy behind which the rearguard were fighting hour by desperate hour. Every disputed yard against the horde pressing us back was to mean the lives of men- our men. Alas, sunset was falling over Greece, and its shadows were black as the shadows of the mountains that blackened the fateful pass.

Greek soldiers were now drifting back fast from Albania, whole groups of them, bloodstained, muddy, ragged men, haggard of face, exhausted in body, the glare of despair in their eyes. Rags were bound about their feet against the pitiless rocks and thorns and mud; their boots were long since worn out. Like ghosts they came staggering down the hillsides, emerged from the gullies, swayed there to stare at us until we beckoned them to come for help and something to eat. Their gratitude at tins of bully beef and army biscuits was pitiable. Many carried single-shot rifles obsolete before the last war; what other little equipment they had was long since out-of-date. Yet these were the fragments of remnants of the men who for so long had held the Albanian border against Italy. And now along the road from the pass came shuffling in tragically growing numbers that most pitiable flotsam of war - women and children, sick, and old men. Little children bravely clinging to their mothers' skirts, babies clutched tight under shawls, bowed bodies wearily ploughing through mud, marching with the growing trucks back, back, back through mud and sleet and rain. Night crept down, smothering that long trail of misery but through the gloom behind again and again burst the dull red flashes of war. Those who sank by the roadside - alas, they stayed there.

We gave them all the bully and biscuits and blankets and help that we could, but their numbers swelled until at last we had nothing but silent sympathy to give.

The little Wog-dog battled on with us throughout all the retreat, very quiet now; well he sensed that this road was tramped by the tragedy of Despair.

And yet, when we could no longer help these people they still had a friendly wave for us, a sad smile. But when they thought we could not notice, their faces were grim and set.

Pulling up for a much needed rest one morning I noticed an old woman pantingly trying to dig a hole beside her cottage wall. She was one of the "too old" ones, too weak to join the throng battling back, back, back before the roar of war now so close behind. She must stay.

With a pity near to tears I dragged myself up to see what she was doing, and to ask if I might help. The Wog-dog ambled silently up to her and gazed up at the poor old soul. She was labouring to dig a hole in the frozen earth in which to hide a few tins of bully beef the troops had thrown her; her food, it meant possible life to her for just a little while longer, food that she felt she must hide from a mighty army.

"Devourie! Devourie!" she smiled and pointed towards the fateful north.

Hastily I dug the hole, dropped the tins in, covered them and made as sure as I possibly could that no one could suspect a hole had been dug there. That morning was bitterly cold, but through lack of sleep and exhaustion I was sweating before the job was finished. I glanced round, and there she was sitting on the edge of a little stone wall nursing Horrie and murmuring down at him smiling at him from a face of a thousand wrinkles. The guns behind us were breaking into a roar of sound.

Propping the spade against the wall, I crossed over to her and taking Horrie put him within my coat. His little head immediately popped out and gazed at the old woman. With tears welling to her eyes she smiled and put one hand to mine and one to Horrie's head, she mumbled something; I gazed down not daring to speak but Horrie licked the old, gnarled hand.

"God protect you!" I muttered, and hurried back to the convoy.

Raid after raid hurtled down on the badly battered convoy, blasting truck after truck in geysers of mud and splintered metal. As the troops staggered back from cover it was to work like maniacs pushing the burning wrecks from the road. There must be no hold up because following convoys and finally the desperately pressed rearguard must fight a way back along this narrow road around which there was no detour. The lives of all behind depended on those in front to keep moving.

When in sight of Larissa we were too numbed from our own bitter experiences to say much. Larissa was in ruins; God help the friendly people who had cheered us as we passed through only a short time ago! Now only a heap of stone was the church and high steeple where the stork had surveyed the peaceful scene below. I wonder what that stork thought when the bombs came screaming down to blast the church and his home and the little town to ruins.

The Germans had given the folk of Larissa three hours to vacate the town and the convoy had halted to allow the people to pour out and get a start on down the road ahead. It was humane, and the only thing to be done, though it was to mean hell for us. A Company was hurriedly turned aside to defend the battered airfield while the rest of us closed up where the head of the convoy had halted at Larissa. A train was there, already hopelessly overcrowded. Numerous people and children were still hurrying towards the station while overhead three enemy planes hovered like birds of prey. When riding through the centre of the town I passed thirty little girls all dressed in white marching quietly towards the railway. Six children walked in front carrying a white sheet quartered by a broad red cross; one child held each corner and one each side. The pathetic little party marched bravely on with the roar of the planes low overhead; several little smiles, several little hands fluttered at me as I passed by; surely such little children carrying the emblem of mercy would have nothing to fear. The little party was in the charge of five nuns. Probably they were from a convent somewhere in Larissa.

The convoy pushed on. Outside Larissa the road south was a weary stretch of flat, swampy ground, a nightmare with rain and mud and the churning of countless vehicles. Over this Devil's Crossing there was not a blade of cover so the convoy pushed stolidly on against the scream of planes, hiss of machine-gun bullets, roar of exploding bombs. Hell was here, night and day. Any truck that slid a yard off the road was finished, any truck hit by a bomb was finished. We plodded slowly on, trucks, soldiers, women, children, old men, wounded men, dogs. The roar of hell was above and bursting among us all the time.

While checking the passing of my section trucks I noticed one missing and with a sinking of the heart I turned back towards

Larissa to try and locate it. All this time the Wog-dog had stayed quiet as a mouse within my greatcoat. I could feel the warmth of him. He knew something awfully tragic was happening.

I passed a shocking sight as I pushed the bike through the mud. I dare not stop a moment lest I break down or go mad. The huddled, muddied bodies of children in white, fragments of a white sheet with torn cross all muddied and red, red, red. The little bodies lay there like crushed, broken, little red and white roses. An Australian ambulance crew was bending over them, but one glance showed me that few indeed were left to remember their little play-mates. I pushed on, hatred rending me, cursing those inhuman devils, those accursed Goerings and their magnificent Luftwaffe.

Blessed for me it was that soon afterwards Horrie gave me a great fright.

To the roar of planes he leapt from the greatcoat with a warning bark and raced for cover. His little legs could carry him faster through the mud than mine. He paused to see that I was following; then, at the whine of falling bombs, barked frantic warning and leapt ahead plainly saying "Every dog for himself!"

He vanished, and the bombs exploded in mud and flame and choking smoke. As the planes roared away I struggled up from the mud and whistled for Horrie. Strange! He did not return. Feeling horribly sick I staggered in the direction I had seen him last, whistling piercingly. His smothered bark made my heart thump again. I found him in a narrow trench, far safer cover than we had - but what an awful sight was Horrie. He had jumped into a latrine and - phew! The boys laughed hysterically as we struggled back to the trucks. It was not until late that night that I could get petrol to wash the little wretch.

The Greek forces on our left had now been overwhelmed and the enemy was pouring into the rear of the Thermopylae position from Yannina.

The convoys ploughed on and on. The rearguard fought their way back yard by yard, their blood staining the mud a slow purple. Was it not off the shores of Greece that the Romans fished for their royal purple, Purple of the Caesars!

It was while we worked like maniacs at a bogged truck that Don came along. I laughed a silly laugh. How very, very glad I was to see Don. The truck's front wheel had stopped, choked with mud wedged tightly between the mudguard and the wheel. Smoke was issuing from the tyres because of friction between tightly packed mud and rubber; we wrenched the mudguard upwards, the truck moved and slowly the convoy pressed on.

"Phew!" said Don as we parked for a breath, "it's not daisies I smell!"

"It's Horrie!" I gasped.

Don looked questioningly as he held his nose.

"If you ask any more questions I'll burst into hysterics!" I yelled.

Don grinned, crawled up on to his bike and rode away on his job.

At last our section crossed that Devil's mud and ploughed on over the better road towards Lamia. Thank God, the wintry sun sank at last!

We pulled off the road while the drivers just stumbled off the trucks and sprawled down to camp for the night, dead beat, utterly done in. I found Ron Baker's truck where I'd arranged to meet Don and the boys. Horrie leapt out of the greatcoat with a yap of delight to meet his numerous friends and ravenously eat the scraps of bully and biscuit showered upon him. He trotted to truck after truck on visiting bent, and stretching his little legs. I was very worried at seeing no sign of Don. I had been riding at the head of the company, he in the rear. Poppa loomed up out of the darkness. I knew he was weary but his alert eyes and grim, set mouth showed he was set for any emergency.

"Where's Don?" he asked.

"Missing."

"We'll find him. Who is the last man to have seen him, and where?"

"Ron Baker. At the turn-off."

"Come on, and be smart about it!"

So Ron Baker and Reg. Jenks and I slouched back along the dark road amongst the strangely quiet trucks to the turn-off. We scouted round, and on hands and knees in the darkness traced where two trucks had missed the turn-off.

"It's all right," sighed Sergeant Poppa. "Don's gone after them. He'll bring them back before they can run into the enemy lines. We'd better return to the unit in case we're needed."

As we ploughed slowly back, several little fires suddenly appeared on the hills around.

"Spies or Fifth Columnists!" growled Poppa. "Lighting fires to show the Luftwaffe where we are." He seized a handy anti-tank gun and blazed towards the fires; the stutter of a machine-gun suddenly awakened the night as it sprayed bullets towards the tell-tale fires. The fires were quickly extinguished.

"They didn't think we'd wake up," growled Poppa. "I'd love to have enough men to surround that hill."

"It's only a very little hill, hardly a pimple," said Reg. "but I doubt if there's a man in the unit capable of climbing it. I can hardly stand myself."

Dead to the world, it was two o'clock in the morning when I was forced back from deepest sleep. The dim form of Poppa was bending over me.

"Your turn for picquet duty, lad," murmured Poppa.

I dragged myself up and staggered into the darkness. I crawled some fifty yards up the hill from the trucks and fought against sleep with my tortured mind going back, back to the little children of Larissa, flitting from there to Don's absence, then back to the little children again. Horrie sat between my legs; I felt the warmth of his little tongue in sympathy upon my numbed hand. All pressed down we were by the pall of nights deathly silence over all the exhausted convoy. Away back - the thunder of guns, crackling bursts of rifle fire. But around our sleeping camp was utter silence.

A low growl from Horrie startled me back to duty.

"What is it, Horrie?" I whispered. His body stiffened, his hair stiffened, he took a few paces up the hill, stopped, and growled again. "The men who lit the fires!" flashed through my mind and I clenched my rifle while sinking to the ground and trying to silhouette the hilltop against the sky. I crept up to Horrie, laid my hand upon him and felt him quiver in warning. He could see what I could not see.

"What is it, Horrie?" I whispered.

He growled menacingly.

Strain as I may, I could only distinguish the dim outline of rocks. Horrie began to advance with a threat in every movement of his tiny body; it was no longer comical to me. Bent low to the ground I followed on noiseless feet. I dare not call to the sleeping men below lest it prove a false alarm but I knew this was no false alarm. Horrie insistently advanced and presently I saw he was growling towards a crouching rock now barely thirty feet away. With the sickening feeling of a bullet in the pit of the stomach I whispered "Stay put, Horrie!" and touched him and then crawled away to outflank that crouching rock, my eye upon it, the rifle poised to fire. I had only gone a little way when Horrie growled and rushed in at the dim form on the ground. I leapt in and called "Halt!" as a shadow rose to Horrie's attack.

"Kalanite!" came the cry from a figure now standing motionless. Growling fiercely, Horrie ran back to me to face back at the stranger with a growl.

"Advance one pace," I ordered, "but no more or I shoot!"

Again he replied "Kalanite!" Kalanite was Greek for "Good night", but the shadow took one pace forward. I moved in close and motioned him down to the camp with the rifle, Horrie growling at his heels. I manoeuvred the stranger to the O.C's truck.

Captain Plumer sent for the Greek interpreter and interrogated the man. His story was that he was a shepherd seeking goats that had strayed from this flock upon the hill. Surprised at sight of the trucks (at night he could not have distinguished them from where he was) and startled by the dog he had laid down, fearful he might be mistaken for a Fifth Columnist.

He stuck to the story, which was feasible enough. In the circumstances, nothing could be done. He was escorted from the camp to return to his flock. When daylight came I searched the hill. There was no sign of the shepherd. Nor was there the faintest sign that animals of any description had been anywhere upon that hill for a very long time.

"You little trimmer," I said to Horrie, "you probably saved the camp. That fellow was in readiness to signal the Luftwaffe. We would have awakened to a tornado of bombs - those of us who did awake."

Horrie views his cobbers from a Vichy French tank, Syria

(Top) Horrie resting on the beach at Gaza; (lower) Horrie at Australian Soldier's Club with Rebels at Tel-Aviv.

(Top)Horrie and Imshi; (lower) Horrie at Gaza.

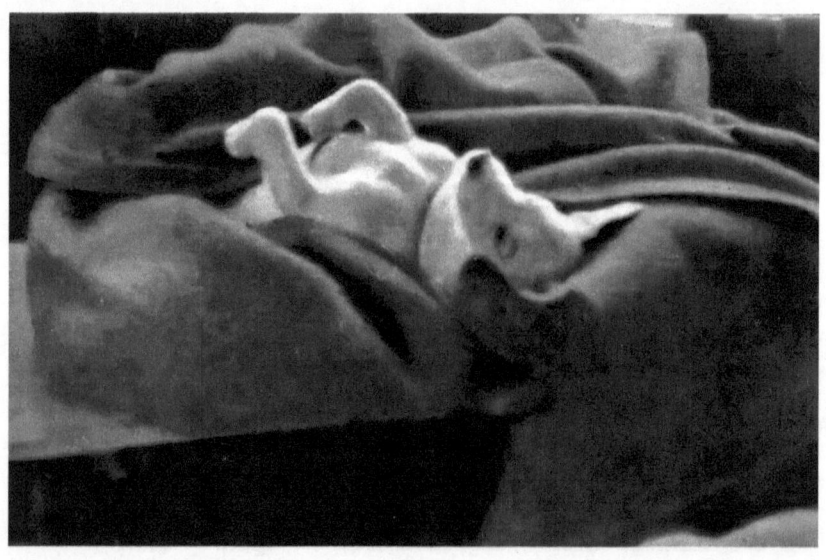

(Top) Horrie aboard H.M.S. Defender (destroyer) taken after being rescued from the Costa Rica; *(lower)* Horrie "bashing the spine" in approved Aussie soldier style.

Horrie (the pup) experiences his first bath just prior to being accepted as the Rebels' mascot, Ikinga Marint, Egypt..

Horrie in his travelling pack, taken shortly before we left Palestine on our return home.

Horrie is actually in this pack, taken at Port Tewfik, Egypt, while we leave on the march to the embarkation point when we boarded the West Point.

(Top) Horrie watching Aussie soldiers playing two-up aboard the Chakla on the way to Greece; *(lower)* Horrie and self asleep during off-duty period in Crete.

(Top) Horrie on self-appointed guard outside our tent, Deir Suneid; (lower) Horrie shows the Rebels that he can sing to the accompaniment of a mouth organ, Khassa, Palestine.

(Top) Horrie shows interest in the music he can hear issuing from the ear-phones of a portable army wireless; *(lower)* Horrie supervises the digging of a slit trench, Palestine.

12

MURCHIE STAGES
A PRIVATE WAR

WE moved on again. The Luftwaffe came again. Bad news came that the Sixth Australian Division had suffered heavily and was being forced back. We knew there were no reserves. We must push on all the faster. One unspoken thought worried everybody's mind. If, happily the enemy failed to break through our rearguard then, eventually, we must reach the sea. What then?

Would we be driven into the sea or could the Navy rescue us?

The nearer we struggled towards the smoke pall glowering over Lamia the fiercer the Luftwaffe attacked. Truck after truck roared up in flames. With the energy of despair the men pushed the wrecks clear and the convoy kept moving on. We hardly spoke now of the rearguard holding back the enemy behind us; we were expecting shells any moment. But the rearguard fought and fought, pressed back yard by yard.

Throughout the anxiety and toil I was very worried over Don. Every now and again I would come across one or other of the Rebels grim and haggard at their different jobs. But no one had seen Don.

Suddenly, quite inexplicably, a mob of cattle appeared among the trucks on the narrow road. The trucks pushed on. Many of the poor cattle were sent crashing down into the ravine below, there was no room on that tortuous track for cattle and the trucks of a bitterly pressed army. Again and again cattle were driven on to the road, again and again we crashed through them. Thus the Fifth Columnists among the shepherds tried to delay us.

We were passing through a small village when a Greek woman stopped our Regimental Sergeant-Major Kelly, talking mysteriously and pointing to a man idly reading a paper beside the road. The curiosity of the woman had been aroused. The R.S.M. walked towards the unconcerned man who suddenly leapt up and ran. The R.S.M. instantly dropped him with a stone. Greek soldiers came running and seized the man. The paper was examined; there was another paper marked by the serial numbers of the trucks, their type and approximate number, the number of soldiers and equipment they contained, a first-class military check-up of everything passing along that section of the road. He was shot on the spot by the Greek soldiers.

Lamia was now densely blanketed in smoke, German planes roaring above the smoke. The road went through the town that was now becoming a ruin of crashed buildings. Here, as often before, we silently witnessed the deathless patriotism of the Greek people. Although their town was in flames, although buildings were crashing around them, groups of Greek men and women were slaving like Trojans to clear the debris from the road so that our trucks could go through. And we were the convoy of a beaten army, an army that had come to save them, an army now abandoning them.

We had come to cheers and song, laughter and joy and flowers. As derelicts we were returning, sullenly rumbling back through the flames of war.

We were gasping as slowly we rolled through the burning town; the dust clouds and choking fumes and great heat made breathing painful. Now and again with a tearing crash a building would come down, spilling masonry across the street and the Greeks would rush the stones to pull them clear of the road. Further through the town the carnage was ghastly; bewildered old people were wandering aimlessly through the broken streets, terrified children wailing over forms lying pitifully still or with feeble arm blindly trying to protect the little ones. Screams of the hurt and dying were mocked by wailing sirens as the Luftwaffe came roaring back again. We had ploughed through hell after hell since the hell of Hellfire Pass and each hell was more anguishing than the last.

The little Wog-dog kept very quiet; he never moved from my coat all that day; I dared not let him free even had there been the time to do so.

And now Murchie was missing as well as Don. I met Poppa for a few hurried minutes; he had a bruised shoulder from a nasty spill when a stone from a toppling building upset his bike.

"It's nothing," he insisted as I stared at his dust-caked, bloody face. What had really happened, as one of the truck drivers told me later, was that he had purposely crashed into a heap of rubble to dodge a terrified child who had run across his path.

"Gordie is O.K.," said Poppa. "Feathers is still going great guns with A Company; Fitz is O.K. too; the others are sure to turn up. No need to ask how Horrie is," he grinned, for Horrie was frantically trying to squeeze out of the greatcoat to embrace the old sergeant.

"Ah, well," he said, "will be seeing you."

And trying to look as if it did not hurt he mounted a commandeered cycle and carried on with his job.

My cycle was smashed, so now I travelled with Ron Baker as relief driver in his truck with Horrie travelling like a gentleman in the cabin.

We camped in open country that night. Luckily we dug a few shallow trenches out from the camp. I could swear I had not been asleep a second when Horrie's frantic barking brought me instinctively to my feet at the run. All the sleep-exhausted men were leaping up at a crouching run, racing to the trenches by instinct. It was just daylight and the Luftwaffe was upon us. Horrie by now was barking agitatedly from the trenches. He vanished as the first bombs whined down with men tumbling into the trenches.

They bombed us ferociously, then put us under extra heavy machine-gun fire and - did not hit a man! One bomb blew a big crater within the camp area and Bert Bottom scrambled into this and opened up with a Bren gun. Tim Overall and Fred Richardson quickly joined him and blazed away at the plentiful targets. They put up a great effort. We cheered when we saw the enemy planes quickly forced to fly higher to dodge the streams of bullets.

The only casualty of that particularly vicious raid was a poor old donkey that had been quietly feeding out from the camp.

You may be sure that Horrie the Wog-dog received lots of pattings and cheerios for his timely warning and more breakfast than even he could put away.

Quickly we were on the road lest the Luftwaffe catch us napping.

As on other days the poor Greek villagers, men and women, waited by the roadside to fill the bomb craters as the Luftwaffe screeched over. Thus the trucks rolled on; the craters were filled in almost as quickly as they were made. On 24 April we were hidden in an olive grove near the little village of Ulanda. For some reason our own unit had received orders to remain hidden throughout the day and be prepared to move out at night. For this few hours' rest we were quietly grateful.

At midday, Poppa and Horrie and I took a stroll through the pretty village.

Poppa's professional eyes fastened on to a battered old truck pulled up in front of a little inn.

"It's been in the wars," I remarked.

"An old-timer too," said Poppa, "and seen a lot of recent action."

"Why it is part burnt out," I exclaimed, "and look! Riddled like a sieve with bullet holes. How on earth did they keep it going!"

"Because they're good men," answered Poppa grimly, "whatever of them are alive."

"I shouldn't think any of them could be judging by the truck."

A roar of laughter startled us. We gazed towards the inn. There came drifting out to us an unsteady voice braying the Sixth Divvy song, "Old Blamey's Boys".

Old Blamey's Boys, (roared the voice),
6th Divvy Boys
Fighting for Victory, Liberty, Democracy,
Oh, Hitler we warn,
We're the A.I.F. reborn;
Like good old Gunga Din
We can take it on the chin
'Cause we're Old Blamey's Boys.

Poppa was staring at me questioningly. With a thrill we stepped straight in to see that "voice".

Yes! It was Murchie, standing unsteadily on a table beating time to the song with a bottle of wine and surrounded by laughing Greek soldiers. Horrie yelped frantically, and struggling from the greatcoat, raced down among the tables.

"The Wog-dog!" yelled Murchie, "the good old Wog-dog! Where's the boys?" Then he caught sight of Poppa and me and we all joined with Horrie in excited greeting.

"Struth!" exclaimed Poppa, "where the hell have you been? Dodging duty as usual, no doubt! But what a sight!"

Murchie wore a ten-day's growth of beard, a dirty face from which Horrie was trying to lick traces of mud, a German officer's cap, a German Luger pistol stuck through one side of his belt and a wicked looking knife through the other. Across his shoulders was slung a battered Tommy gun.

"Horrie the Wog-dog!" laughed Murchie and held the excited Horrie overhead for all the Greek soldiers to admire. I noticed Poppa's glance and for the first time saw that two Kiwi (New Zealand) soldiers were supporting Murchie. They looked just as much like cut-throat bushrangers as Murchie did and were armed very similarly. These obviously were the owners of the battered truck.

"Looks as if they've been waging a private war of their own up in them thar hills!" grinned Poppa. "I wonder where their poor mates are."

Poppa knew and I knew. If there had been any mates left alive they would have been with them here.

"Oh, I forgot," laughed Murchie. "Meet Archie and Bash, cobbers of mine. Poppa and Jim and Horrie the Wog-dog!" We shook hands with the New Zealanders while Horrie insisted on being introduced in his own doggie way.

"Wine! Wine for the mob, George!" shouted Murchie and waved his arm. They all made room at the table for the two Kiwis and us and Horrie; the ragged Greek soldiers closed around us, laughing and joking. George came smiling with his arms full of bottles, glasses were filled glasses clicked. Murchie pulled an expensive gold wrislet watch from his pocket and with the flourish of a millionaire handed it to the innkeeper.

"Cut this out in wine for the boys, George," ordered Murchie.

Poppa gasped at sight of that watch.

"Wait a bit, Murchie," I protested. "Hang on to the watch. Poppa and I can buy the drinks."

"It's all right!" said Murchie with a contemptuous flip at the watch now entering the innkeeper's pocket, "it's all right; plenty more where that came from. German officer gave it to me!"

The swarthy innkeeper grinned broadly, the New Zealanders laughed, the Greek soldiers roared as at a priceless joke.

"Hang on to the dashed thing!" exclaimed Poppa angrily. "You won't get another like it in a hurry!"

"Jush where you're wrong ash usual, old war-horsh!" frowned Murchie triumphantly. "I'll show you! Here's 'nother!" From his tunic he produced another lovely watch. Poppa gasped, I stared.

"German officers very kind!" declared Murchie. The Greeks roared.

"They must have got mixed up somewhere in a lively ambush," whispered Poppa, "and the Germans got the worst of it."

"Here George," yelled Murchie, "glass-top table too cold for Wog-dog to sit on. Bring cushion or something, and jug of milk!" Murchie went through the pantomime of milking a goat.

Between numerous glasses of wine we learned by degrees of Murchie's adventures. During a particularly heavy raid he'd returned from shelter to find his own truck burning and the enemy on top of them. He'd leapt aboard a Kiwi truck and his adventures started properly. He was mixed up in continuous fighting and retreat, in which it increasingly happened that small groups of men became separated and fought on their own. The main body of New Zealanders had been pretty badly knocked about in the confusion. Murchie's truckload of diehards seemed to have been in the thick of it. When casualties occurred they were quickly replaced by Greek soldiers.

"Plenty reinforcements!" declared Murchie. "Plenty. Good fighting boys, too. Fought like wildcats. Look at my warriors here!" and he swept his arm towards the grinning Greeks. They certainly did look as if they could fight like wildcats, and cut a throat with the best of them. Some of them wore bloody bandages. Their greatest trouble, it appeared, was to keep the old truck going.

"There was plenty of petrol," declared Murchie. "Trucks were knocked out all over the place. It was the mechanical part that counted, just to keep the old jigger jigging in between scraps. But Archie and Bash can keep any jigger jigging, no matter if the whole German army are at their tails."

And Archie and Bash grinned at the compliment.

They had fought their way through, then followed the convoy south and eventually found their way to Uandra, delighted to find now that part of our own unit was stationed not half a mile from the cafe.

"What will you do with your army now?" inquired Poppa.

"Oh," declared Murchie grandly, "they'll come in handy if the evacuation becomes 'unstuck'."

Our hearts sank. So it was evacuation. Murchie had voiced what we all really knew must take place. But Murchie had been fighting amongst many remnants of mixed brigades who realized the position beyond doubt.

"Just how do you mean?" demanded Poppa.

"Why," answered Murchie, "we'll mobilize the Greeks into guerrilla bands. There are plenty of them willing too, remnants of their Army. It's all smashed up. We'll mobilize them and hang out in the hills until we find some way of getting back to Egypt. There must be plenty of small fishing craft around the coast we could commandeer if we become sick of fighting, which we won't!"

Poppa and I agreed that this was after all the logical thing to do if it came to the very worst. As a matter of fact, it was the very thing quite a number of our lads did when they found themselves cut off during the evacuation.

We had to tell Archie and Bash all about the little Wog-dog's adventures and just didn't he take all the petting that came to him. "These blokes are good blokes!" he wagged across the table to me. Just then a Greek soldier came running in the door and straight up to our table.

"Turkey!" he shouted excitedly. "Turkey bom-bom German!"

Turkey at war with Germany!

In an instant Greeks sprang from chairs; the inn was crowded; we could hardly hear ourselves talk. What good news!

"Is it possible that Turkey could cut the German line-of-communication between Albania, Yugoslavia, and Greece?" asked Archie.

"If only she could!" replied Poppa. "And if the British could hurry more troops and equipment across here to us we might yet get the enemy between two fires!"

"But," I protested, "aren't we too late? The Greeks have already capitulated to the Huns."

"To what?" came a shout from the door as Horrie leapt from the table with frantic bark.

"Don!" we shouted.

13

THE BITTER RETREAT

WE crowded around him; he was filthy, ragged, caked with mud through which grinned the old Don smile. A wonderful reunion. To hide his feelings Poppa exploded, "So the prodigal sons have returned - and about time too!" as Murchie and Don shook hands.

"All right, all right, take it easy!" laughed Don as he grabbed the excited Horrie. We somehow found room at the table and filled another glass, the signal to fill all glasses.

"Swig it down, then start talking," hoarsely advised Poppa.

"Introduce me to your friends first," laughed Don.

We did so with gusto and Don settled himself for his yarn; it was wonderful to thus meet one by one and hear the dear old Rebels again.

"First of all," smiled Don, "Feathers is O.K. At the present moment he's asleep in one of the trucks, dead beat. Fitz and Gordie are also O.K.; with a little luck they will pull in here before we leave tonight. How's that for news?"

"Hooray!" shouted Poppa. "Let us get tight!" Poppa was already nearly tight and trying to hide his joy that all his old Rebels were coming together again.

"What on earth happened to you?" I asked.

"Well, after we got over that bad patch outside Larissa the old bike went on strike." And Don carried on with one of those many romances of the mountains and the passes, of attacking and retreating armies, and cut-off groups of fighting men in which only the lucky ones came through. Don had joined up with New Zealanders who had circled back to the enemy and then escaped back to us with the last of the rearguard. I'd absorbed quite a lot of wine by this time and my entry of Don's adventures in my diary, I found later, was very incomplete.

When Don had finished his yarn the shadows of evening had fallen.

"What about your outfit?" I asked Archie and Bash.

"Guess they're split up all over the place," Bash replied. "How about if we tag along with you until we locate them again?"

"That's O.K.," answered Sergeant Poppa with military dignity. "You men of the New Zealand Army are on the strength!"

"Have you heard anything about Turkey being at war with Germany?" I asked Don.

"Plenty of rumours," he answered, "but we do not know the truth."

"How about your Greek commandos?" demanded Poppa of Murchie.

"Dash good fighting men!" declared Murchie with a thump at the table. "Where I go they go!"

"Thash all right by me!" declared Poppa. "We'll put them on the strength."

"They can have the truck," said Murchie, "and follow us when we pull out tonight."

"Thash decided!" declared Poppa and rose unsteadily. "And now we must go back to camp. There'll be something doing tonight!" he added darkly. "Give orders to your Greeks to pile into the truck and follow ush."

So we filed back to the lines with the truck loaded and packed with Greek soldiers. We parked them at the rear of our lines and then filled the battered old truck with rations. Shortly after dark the convoy moved stealthily out on to the road.

"German air-borne troops are expected to land at Corinth and cut off our retreat by blowing up the Canal bridge," Sergeant Poppa told us shortly after we had started, "so the convoy keeps moving. See to it!"

And the race commenced to get the convoy through before daylight. Every second or so headlights were thrown on recklessly. It was a "do-or-die" stunt. Hundreds and hundreds of headlights lit the mountain levels as we raced towards Corinth. We made it, passed through Corinth and raced south for Argos. The paratroops were dropped too late. Just after we'd passed Corinth they came to ground in early morning and blew up the main bridge to Athens. The convoy missed disaster or at best a German prison camp by a matter of minutes.

Just at daylight our weary, thankful unit pulled into a friendly little olive grove and there hid for the day.

"Where are my commandos?" demanded Murchie immediately we piled out of the truck.

We searched for the truck, but the Greeks were not there.

"I'm afraid the old truck must have broken down," said Archie with a sigh. "I felt certain when we were travelling it could not have survived that mad run last night."

It was a bitter blow to Murchie; he and the Kiwis and Greeks had been through a lot together.

We found Feathers, but failed to locate Gordie or Fitz, though they must have been somewhere near.

"What day is it?" asked Poppa suddenly.

We had no idea.

"Anzac Day! April 25th!" he declared.

"The landing on Gallipoli!" I exclaimed.

And Poppa was launched on the Landing. We listened sympathetically to the old warrior's memories. Old Sergeant Poppa was in serious vein. We felt serious enough, heaven knows; this meant another Gallipoli for us - not such a creditable one.

"Let's wander up through the village," suggested Murchie.

"Excuse for a bottle of wine," growled Poppa.

"Yes," replied Murchie, "I saw your tongue hanging out."

We passed many trucks hidden among the olive trees with tired soldiers digging shallow shelters for protection. The enemy planes would be seeking us any moment now. Horrie scampered around investigating bushes and any thing that looked like a tree to race back to us full of beans and fun.

"He can take it!" nodded Poppa gloomily.

"Yes, he's setting an example," said Murchie brightly.

But for once Poppa did not "bite"; he was taking Anzac Day very seriously. To spend this day hiding and running away from the enemy vividly contrasted against that great Landing at Gallipoli.

"Cheer up, Poppa," I advised. "It's not over yet."

"No," he replied, "but it stinks a bit."

"I'd rather stop with the Greeks than run away," frowned Murchie.

We walked on quietly for a while thinking of the game little country and the old people and children we were leaving behind.

"Perhaps we may come back some day," remarked Don.

"I'll be in on that!" declared Feathers.

"Me too!" replied all of us and the Kiwis.

As we strolled into the village we saw the Greek folk in the street quietly talking in little groups. Poor souls, they knew what this was to mean to them. Even so, they had a smile for us as we passed by.

"For God's sake, let's get a drink!" growled Poppa.

We found a little deserted cafe. Eventually the proprietor came in and served us with wine. Presently, Horrie's energetic barking outside brought us to the door. Horrie, while investigating a little cottage on the other side of the street, startled a very annoyed hen and her brood.

"Horrie, get back!" I called, and very reluctantly the little wretch came to heel and followed us back into the cafe.

"Better leave the commandeering to the enemy, Horrie," grinned Don.

"He thinks he may as well snatch a feather or two before they arrive," said Poppa grimly.

Some time after we had settled down again at the table, an elderly Greek woman timidly approached, smiling wist-fully. She was carrying two trussed up fowls. She held them out to me in voiceless invitation while Don held the eager Horrie.

"For you, for you!" she then said in hesitating English.

Poppa thrust a handful of money into her protesting hand; she was voluble and almost violent about it but we firmly insisted. Regretfully she put the money away and smiling, beckoned us to follow her, pointing through the open doorway at her cottage across the street, by signs inviting us to a drink and bite to eat.

"I'm going to accept," growled Poppa. "Come along."

With Don firmly carrying Horrie we followed her across the road to her humble home, a little square cottage of rough-hewn stone with an archway covered with grape-vines, looking clean and cool against the background of whitewashed stone wall. Beckoning us into the dining-room she motioned us to be seated at a large table. The neatly clean room seemed to occupy most of the cottage. The furniture was home-made, rough but comfortable. On the mantelpiece above the open fireplace stood a little bronze crucifix. The room was bright and peaceful but there stole upon me a most uncomfortable feeling of sorrow. I noticed that the Rebels, when they did speak, did so almost in whispers. The old sergeant sat glum and frowning, Don spoke to Horrie in low voice, Feathers and the Kiwis sat quietly. The spell was happily broken when the old lady returned from the kitchen with a bottle and glasses, brown bread and white, and sour cheese made from goat's milk. She smiled to us to help ourselves and put a little glass bowl with gherkins in olive oil upon the table.

We set to, smiling our thanks. Her kind face beamed as we made short work of the little offering; it really was tempting, but I realized Poppa was eating like a wolf just because it so obviously pleased her as she fussed around us.

When we protested that we simply could not eat one crumb more we sat and tried to make conversation for a while, feeling queerly embarrassed and wishing we were back in the cafe again. She smiled, nodded "wait a moment" and disappeared soon to return with a small photo of two Greek soldiers, proudly erect in their national uniform. We saw at a glance they were father and son.

We gazed at it each in turn. Standing quietly there her face reflected pride and sorrow. Not knowing what to say, we handed the photo back. She closed her eyes and for a moment inclined her head on her hand. And then, through tear-dimmed eyes, she smiled at us and whispered "Italiano. Albania."

Father and son slept in Albania.

Poppa stood up and kissed her hand. "Let us get out of this!" he whispered.

We made our adieux and silently walked down the narrow winding street in the direction of camp. We heard her voice calling to us; she was hurrying after us holding up the trussed fowls that I had hidden in the room. We had to take them again. We tried to apologize for being so stupid as to forget them.

Back at camp, we roasted the fowls over a slow fire.

"Anzac Day!" said Poppa bitterly.

It was not a happy meal, but Horrie enjoyed his share and more.

After dark, while the camp was quietly preparing to move, the Rebels got busy. Scrounging all the blankets and rations we could we sneaked away up into the quiet, dark little street. Very quietly we crept to the cottage of sorrow and loneliness and gently placed the bundle under the vine-covered arch. Then we sneaked away.

The convoy started on its last run; in the early morning we arrived at Kalamalta and feverish activity as convoy after convoy came pouring in followed by war-weary troops - British Tommies, New Zealanders, Australians, Greeks, Yugoslavs. For miles around us troops were wearily plodding to positions, artillery moving to their last position, weary movement everywhere. For here was to be staged the last organized stand in Greece.

The oncoming enemy had to be held back while all troops possible were to be evacuated - if possible!

As for us of the convoys, we broke into an orgy of destruction. Thousands upon thousands of trucks were destroyed, the oil and water drained from the engines then the motors raced dry until they seized; tyres were slashed and all conceivable damage done except burning as the fires would have betrayed the positions to enemy planes. Now it all depended on the British Navy. If the Navy could not evacuate us that night then - it would be just too bad!

Blankets and goods and all stores possible were given to the frugal villagers, old men, women and children who came swarming to us for many miles around. Poor folk. It all depended on the food they could hide as to whether they would survive the fast-coming bitter months.

All day long giant Sunderlands were roaring in hastily to evacuate the many wounded from the beaches. It was uncanny how again and again they missed the Luftwaffe; the Luftwaffe would roar in to bomb the countryside, then hurry away for more bombs and ammunition. By magic the Sunderlands would appear, quickly load with wounded and be away just as the Luftwaffe reappeared again to bomb and machine-gun position after position all through the tragic day.

Greek villagers for miles around were guiding weary troops to places of concealment until the anxiously awaited night should fall, a night that seemed never, never to be coming. While men and women still toiled on the road filling in bomb craters as the last convoys struggled in, weeping girls clung to soldiers imploring them to take them away before the Germans arrived.

And right here, amongst the slaving toil of the Rebels out of the corner of my eye I noticed a little romance. Two beautiful young Greek girls. In no time (I could not take my eyes off them) they were in Australian uniform, toiling among all the muddy, oily, sweaty lot of us.

Many hours later I noticed them on the transport as through that hellish night we steamed for Crete. Alas, since then we have wondered whether they jumped from the frying-pan into the fire.

But we were still on the land fronting the beach, longing for the night that would not come. As we toiled at destroying the trucks, Greek families with their men folk missing, packed our goods on donkeys and then on their backs to hurry the good food and warm blankets away to hiding places. Old men, women and children toiled with the energy of despair. Well they knew what was coming on the swift wings of flame and rapine and murder. Pray God a similar tragedy never happens to my own dear country!

The Rebels, like all others, toiled to exhaustion by the roadside. We watched out for dear old friends among which were cobbers of the 1st Anti-tank Regiment. We located a few, but others were never to be seen again.

A war-stained Digger in a crowded truck noticed Horrie and yelled "Stick to him, Dig!"

"My - oath!" shouted Don as the truck hurried past.

The faces of the Diggers under their mud and sweat were bitter in defeat. Throughout it all I sensed everyone had learned a lesson in courage from the quiet Greek people.

Darkness found us crowded on the beach. A great crowd, crowds of silent men, silent shadows breathing with the night a voiceless prayer of bitterness and despair and wistful hope.

Would the Navy make it?

Before us were the gods of war belching flame with rending crash of bomb and shell, roar of rifle-fire. Behind us, at our very feet - the sea.

Would the Navy make it?

At half-past one that night lights suddenly blinked out at sea.

"Christ!" whispered a voice, "they've made it!"

I near crushed Horrie under my arm; I could sense the feeling of utter relief that arose all along that crowded beach. Sergeant Poppa crouched as if about to leap into the sea, his eyes glaring in the starlight.

"Of course they made it!" he hissed. "You damn old nitwits, of course the Navy made it! They always make it!"

Poppa was in charge of the Rebels and our New Zealand cobbers, Archie and Bash.

Dim shapes came swift and stealthily to beach and wharves, clouds of shadows moved towards them. In silence and mob discipline we swarmed aboard craft that swiftly glided away to troopships to return again and again and again. It was a magnificent job by the Navy and merchant seamen. As for the Rebels, there was no need to hide Horrie now. I just kept a tight grip on him; he was a veteran now. Long since he had learned how to behave from the example of the troops. We piled aboard the destroyer Defender and were swiftly ferried to the troopship Costa Rica. A sad farewell to Greece. Morning found the troopships steaming line abreast, the Costa Rica on the left flank, the City of London in the centre, the Delwarra on the right flank. Our particular escort comprised the destroyers Defender, Hereward, and Hero, with the "ack-ack" cruiser Calcutta.

The Stukas came with the dawn and attacked in screeching waves, hurtling down through the shell-bursts from the destroyers. It did the heart good to see the swarms of soldiers quietly, setting up Bren guns, others taking positions for rifle-fire entirely on their own. This was a heaven-sent chance to hit back. I saw Don blazing away among the crowd, so tied Horrie up and joined him.

"Pool your ammo., boys!" someone shouted, and soldiers appeared carrying large dishes. Into these was thrown ammunition by man after man who had bullets to spare, while the dishes were carried round and soldiers with no ammunition helped themselves. A growing volume of small-arms fire arose from the crowded ships while the destroyers blazed away.

14

HORRIE SURVIVES SHIPWRECK

THE planes hustled down attacking the middle ship now obscured by smoke, drenched by waterspouts. The planes flattened out over the masts, their machine-guns blazing, but they had dived into a hail of rifle-fire and one machine hurtled straight on with a terrifying crash into the sea. A roar of "You beauty burst out above the rifle-fire while cheering drifted across to us from the attacked ship.

"Look out, they're coming!" someone yelled as planes came hurtling at us into a hail of lead from which each in turn violently swerved as its bomb fell harmlessly into the water. The row was deafening as attack after attack developed; even revolvers and anti-tank guns were fired against those gleaming demons.

"Whacko," a lad shouted, "see that one slew off! The stokers did it - they're throwing lumps of coal up the funnel."

In the laughter I saw chips flying from the mast while even the wire rigging was frayed and nicked by bullets. The gamest man must have been the man away up there in the crow's-nest, staring at the onrushing planes while shouting advice and waving his arms in encouragement to the gunners below. Plane after plane came diving apparently straight at him. He had no time to be thrilled because everything was happening too quickly. High on that swaying mast he stood, a hurricane of bullets whistling up from the decks passed by bullets hissing down from the planes. The wind from the screeching machines waved back his hair.

On our decks were packed four thousand men shooting upwards, when above the din the piano in the saloon broke out in a rousing tune, a hearty chorus with the Wog-dog energetically barking approval. Thus the boys who had lost their rifles cheered us on with piano and song. A terrifying roar drowned all sound as another Stuka crashed violently into the sea - a lovely, terrible sight. Then came a gust of laughter at the Wog-dog up on the sun-deck, excitedly barking at the waves swallowing the plane.

The hot reception drove the enemy from us and the planes again concentrated on the centre ship which was blotted from sight by spray from a stick of bombs. But the fire there also was too hot and the enemy bombs exploded harmlessly in the water. For hour after hour the planes came in screeching attack until by midday we had shot five down into the sea. A Lucas lamp from the bridge of the centre ship flashed us the signal:

"Congratulations on the volume of small-arms fire. Good show - keep it up!" And we did - with the destroyers at full speed concentrated furies of sound.

When the bright sun was high in the sky the enemy got our ship. The planes came like comets hurtling out of the glare and we could not see them until they flattened out directly above the mast. Even then they failed to hit us but one bomb screeching down raised the hair on my head. It burst overside with an explosion that rocked the ship while a mountain of water belched up over the port quarter. The rigging came clattering down. Had that bomb hit the deck it would have turned us into a charnel-house. Our ship had fought the good fight but in that last split-second decision made one little mistake - we had zigged when we should have zagged. The engines stopped, the firing ceased, an uncanny silence enveloped us. The plane had vanished; it was their last bomb, very fortunately for us.

The troops remained quietly at their posts. The destroyer Defender came racing beside us and a voice called through a megaphone:

"What is the matter?"

"We've run out of petrol!" yelled a soldier.

Above the laughter came a quiet voice from our bridge.

"Ship badly holed. Six feet of water in the engine-room. Engine moved off its mounting."

"We'll take off the troops!" came from the destroyer. Immediate, quiet action followed as the destroyer came alongside. The ship was to be abandoned. In an instant, I was jumping along the forecastle deck, trying to get to Horrie. Only then I noticed the alarming slant of the crowded decks. Already boats were being lowered, rafts thrown overboard. With considerable difficulty I reached the worried Wog-dog.

"Your first shipwreck, Horrie," I said consolingly, as I picked him up, "but we'll see it through, never fear." He wagged that apology of a tail and tried to lick my face as we slid rather than walked down that rapidly growing list to starboard. Troops already were scrambling down on to a destroyer. I dared not let myself think what would happen should the enemy planes return. From the sun-deck to the destroyer was a twenty-foot drop; troops were sliding down ropes while the little destroyer below was sickeningly rising and falling with the swell. The big troopship at times seemed to be leaning right over the destroyer - she was sinking rapidly. I'd discarded my greatcoat and had nowhere to put Horrie and leave my hands free.

"Catch my dog!" I yelled down to the destroyer.

"O.K., Dig! Let him go."

And Horrie dropped through the air towards grinning faces and upheld hands. He twisted and turned but was expertly caught, his little face turning up towards me. Laughing at his fright, I slid down a rope and was grabbed by those below while Horrie tried to lick my face off. I put him out of the way in one of the destroyer's lifeboats then hurried to give a hand with the ropes, down which troops were swarming like lines of monkeys.

In the nick of time I noticed a lifeboat coming down from the troopship; one end was slipping as the destroyer was beginning to rise from the swell. I leaped towards Horrie's lifeboat and snatched him out just as the two boats smashed together. It was a close call for Horrie.

In less than half an hour every man had been taken from the stricken ship while the destroyers were circling the water, picking up swimmers and men clinging to rafts and lifeboats. Then we were steaming full speed for Crete. We wondered why the enemy planes had not returned while we were helpless; probably their base had run out of bombs. They did not catch us up again until we were landing at Suda Bay, Crete. We had lost our rifles on the ship but the destroyers kept the planes off with a terrific barrage while we leaped ashore and ran for it. Horrie was so pleased at being on solid earth he did not race ahead for cover but ran with me, barking and leaping at my flying legs, urged by machine-gun bullets whipping up the ground. We dived into a cutting and gasped for breath while the planes roared past. I was stroking Horrie when the sound of running feet and panting warned me to dodge two hurtling bodies.

"Made it!" gasped Poppa's triumphant voice above Horrie's frantic barking.

"Are you hurt?" gasped Don.

"No. For the love of Mike, don't say this is Horrie!" Horrie was kissing Poppa and Don in turn.

"I thought you'd make it," laughed Don as he petted Horrie.

"Yes, with a little luck. Where are the boys?"

"We don't know, but think and hope they're all right. What with that night on the beach and the crowd and the hurried embarkation and sunken ships, boys in all units have been scattered everywhere."

"There's a transit depot established somewhere," said Poppa. "We'd better set out and find it."

We found it among the hills, just outside the town of Canea. The troops were rolling in in two and threes, and groups of twenties and hundreds. They were a sorry sight; some were practically naked, many were without boots, some appeared to be ragged half soldiers, half sailors. These were the men from sunken ships. Those whose ships had been sunk under them had lost their rifles. Under an olive-tree we enjoyed our first good meal and cup of tea for many days.

"Horrie is eating like a horse," said Don. "You must have starved him."

"The sailors on the destroyer gave the little wretch plenty!" I protested.

But Horrie raced away with wagging tail and frantic barks.

"Hooray!" shouted Poppa. "Feathers and Gordie and Fitz and Murchie! Surely now the devil does look after his own."

It was a happy little party of Rebels that sat down under the olive-trees and feasted and yarned. We'd all had our adventures - plenty. But just weren't they glad to see Horrie.

"If anything happened to him it would mean the end of the Rebels," laughed Feathers.

"The end of the war!" declared Poppa. "The troops simply could not get along without Horrie the Wog-dog!" and he patted the little fellow's head, the quaint little dog as usual lapping up the petting.

"I notice you're admiring our fashion plate," grinned Murchie.

I had been smiling at Feathers. How he did it I don't know but despite all he'd gone through he was moderately neat and tidy - just moderately.

"I think I'd better scrounge a bit more tucker for Horrie," I said, "while the going's good. Keep an eye on him.

I strolled to the cookhouse and the cook was obliging.

While I was returning to the boys, a voice yelled "Hey, Jim, I've got your dog!"

In surprise I turned round to see Les Jeffers walking towards me; he'd joined our unit shortly before we left Egypt for Greece, but his was a different job to ours and we hadn't seen much of him. I wondered what on earth he meant about Horrie. I thought it was some joke.

"I saw the little fellow looking terribly alone in the confusion when the Costa Rica was going down," he said, "so I picked him up and managed to get him aboard a destroyer. I've got him over here; I felt sure I'd run across some of you boys when events sorted themselves out."

I followed him across to some olive-trees, feeling very curious.

"There he is, or rather, she," said Les triumphantly.

For a second I was too astounded to speak. Tied up under a tree, a picture of misery, appeared the living image of Horrie. But in a flash I noted the slightly different expression, the long tail, the ears not quite so perkily pricked and she was "a lady dog".

"Well I'm blessed!" I exclaimed.

"What's the matter?" Les asked, for the little dog lay there gazing at us very timidly.

"It's not Horrie," I said, "but it's almost the dead spit of him; the same breed too, and I used to think there was not another dog in the world like Horrie."

"I'm sorry if I've taken someone else's dog," said Les.

"You haven't; she would have gone down with the ship if you hadn't rescued her. I wonder if she belonged to one of the crew; they were fearfully busy getting us off the ship before she sank. Perhaps her owner was hurt."

"Maybe that's it," said Les.

"Well look here, you'll have all your time taken up looking after yourself. I've an idea we're going to be kept pretty busy on this island. What if I take the dog to the Rebels? There are a crowd of us to look after her until we locate the crew of the ship. They came off in the destroyer too."

"Good-oh," said Les, and I knew he was relieved.

It was the funniest thing out, the meeting of Horrie and Horrietta. Horrie at his fullest tiny height - with stub tail stiff as a ramrod, showing off his paces to the bashful girl dog. Horrie doing all the manly stuff, insisting upon winning her confidence, and then with tail erect trotting away to show her round the place. After just the right amount of masterly coaxing she followed meekly and obediently.

"So that's that!" declared Poppa. "Are you fellows going to set up a menagerie?"

"She's a new addition to the two we started with," grinned Murchie.

"If you mean that for me, my lad - " began Poppa, but the returning planes cut argument short. We dived for cover in time to see Horrie leading the terrified Horrietta into a foxhole.

The countryside around Canea was delightful, with groves of olive-trees and vines and little farms among hills overlooked by mountains.

The island of Crete lies across the eastern Mediterranean as a mountain mass about one hundred and seventy miles long and approximately twenty wide. All land possible is cultivated with numerous olive groves, barley fields, and grape-vines. The frugal Cretans have even levelled terraces out of the sides of the hills which appear like enormous steps covered with a green carpet of vine. The folk were very friendly, and though their island was not well stocked with food they offered us the luxuries of fruit and eggs and wine - meals fit for a king, especially after the hard biscuits and bully beef which had been our fare for so long.

But that this peaceful scene was to be peaceful no longer was hourly obvious. The bombing attacks grew in intensity, mild forerunners of the terrible fighting soon to take place before the enemy overwhelmed Crete.

Horrie and his girl friend got along famously together, and were rarely separated. In the bustle of preparing for defence there were no such events as ordinary parades, but at such times as the battalion marched out on a job Horrie and Horrietta trotted at the head of the column looking very important. But the bombing terrified Horrietta; she was pathetically frightened, trembling violently for long after each raid despite all we could do to soothe her. She had obviously been very well looked after aboard ship and no doubt at each alarm had been rushed down below to some quiet hideout. But here was the open earth, and the crash of the bombs nearly paralysed her with terror.

"If we can't do anything with her," said Poppa at last, "it would be better to put her out of her misery."

Kind hearted old Poppa! We might be forced to do it, might have to draw lots as to who must shoot her. We would not see any of the ship's crew again; we'd heard they were immediately shipped to Egypt to man another ship.

"The Germans are going to land here," said Fitz, "and the fighting will be hellish. Sooner or later we'll have to do something about her."

"Perhaps we might get some Cretan family to take her," said Gordie hopefully.

"That's it!" exclaimed Poppa. "All hands keep an eye out for some Cretan family that's willing and will be kind to a dog."

But Horrietta simply could not get used to the bombing. During the violent tremors Horrie tried his best to soothe her, licking her face and prancing around her and trying to coax her to lie still. Again and again he'd lead her to a foxhole shelter of his very own, trying to explain that she had only to run and dive in on the first sound of the planes.

But Horrietta continued to live in misery during the air raids.

15

THE WOG-DOG TO THE RESCUE

ONE day brought us a pleasant surprise - a visit from Archie and Bash, Murchie's New Zealand commandos. They had located their own battalion which had camped quite close to us. We enjoyed a great yarn, Murchie especially reminiscing with the Kiwis about their bushranging adventures with the Greeks. Succeeding days, when duty allowed, often found the two New Zealanders yarning with the Rebels. Alas, in the desperate fighting soon to rage around Maleme Airfield, both Archie and Bash were killed.

By this time Horrietta had fallen for Horrie in a big way; she was his shadow, and gazed at him with gloating eyes while he proudly strutted before her. Though a quiet, timid little thing still she was spiced with a touch of vixenish ways.

"She deliberately tries to make him jealous," grinned Gordie one day when the innocent Horrietta did her stuff and was the centre of attraction. With mild, bashful eyes she was sitting back on her tail, a trick the ship's crew had evidently taught her. She was a little adept at sitting up and would not move until we had petted her and ordered her down again. Horrie stalked round and round her with cock-eyed interest, trying hard to imitate her. But his efforts only brought discreet grins; we daren't laugh aloud. For Horrie's tail was too short, he simply could not sit back for his long body overbalanced every time. Horrietta, on the other hand, could get a firm purchase with her long tail.

"You're not in the race, Horrie," grinned Murchie. And Horrie, being a wise dog desisted from efforts which only made him ridiculous.

In cute little ways Horrietta let us understand she too liked her share - and more - of petting. Demurely she would gaze at the ground when Horrie's growl warned she was claiming too much of it.

One evening Poppa returned to camp looking somewhat downcast. "We've got to hand in our rifles," he growled.

Poppa had boasted that among so many weaponless men his Rebels still had their rifles despite shipwreck or anything else. Now we were asked to hand them in, for every weapon was urgently needed. Our full-time job would be signalling.

"'Don' Company are all right," said Poppa enviously. "They landed without shipwreck and their machine-guns and rifles are intact. They've got to fight it out with the rest of the boys. But signal units poorly armed have to concentrate on signalling duties. An air-borne attack on Crete is imminent. We're moving soon. our job will be to keep communication between small striking forces and company headquarters.

"That's serious," I said. "There must be some thousands of men here now with no equipment at all."

"Much worse than that," replied Poppa. "We cannot replace the lost material. Every possible man will be desperately needed soon, but every man without arms will be an encumbrance."

Next morning the Luftwaffe heavily attacked the shipping in Suda Bay. Horrie scooted for cover, barking to Horrietta to follow. But she trembled so violently that she could not move.

"We must get rid of her," said Poppa shortly. "Today is your last chance to find her a home. We leave camp for action stations tomorrow."

We located in a little Cretan home among the hills some kindly folk who immediately took to Horrietta. The ship-wrecked pup would have a good home. We did not let Horrie know her whereabouts lest he should go A.W.L.

It was well that we solved the problem, for food became very scarce. Horrie was to lose a lot of weight; he was from now on to scrounge for his tucker, to sit patiently in front of soldier after soldier, receiving any small portion of biscuit or bully beef the donor could spare. Our rations were cut to the bone. But he adapted himself so willingly to hardships and was always so bright and contented that his show of "guts" helped lighten our own troubles.

Our new quarters overlooked Suda Bay. Small bands of our troops were occupying a steep rocky hill overlooking the bay and their duty was to watch for and deal with any paratroops that soon were expected to attempt a landing. Our particular job was to keep communication between these scattered patrols and headquarters at the bottom of the hill. By daylight we signalled in Morse, using as a flag a white singlet tied to a stick. But at night we had neither signalling lamps nor even torches. We dare not have used them, anyway, for fear of spies and fifth columnists. And yet a vital message might be received at any moment. Important messages were arriving all through the night. Our only means of getting a message through to headquarters was by runner down a long, treacherously steep and rocky hill. A slow method when minutes might prove vital.

Horrie solved it; he became our willing runner for night messages and could get down that hill much faster than any man. By day Horrie and I worked with the patrol that occupied the hill immediately above H.Q. Just before dark I would leave Horrie with the patrol and scramble away down the hill to the hollow olive-tree where I slept near H.Q. If the patrol received a message from distant patrols it was written down, securely tied in a handkerchief and fastened to Horrie's collar. And away he would scamper down the hill. I'd wake up to feel him urgently licking my face and the message was immediately delivered to head-quarters. To let the patrol far above know the message had been delivered, Poppa would fire two rapid shots airwards with his revolver.

Horrie delivered quite a number of messages, no matter how dark or stormy the night. Not one message entrusted to him ever went astray.

A day came when we were stunned by the news, even though we were partly expecting it, of another evacuation. All those troops without equipment, all British and Imperial troops whose equipment had been lost were to be evacuated.

But for the unavoidable loss of equipment of those thousands of men a different story might have been told of Crete. But imagine what we felt at not only having to leave the Greeks behind, but our own mates as well.

I packed Horrie into an empty ammunition-box, for there was no foreseeing how troops might become separated in a ticklish operation like this. Some hostile officer might spot Horrie and order the dog be left behind.

The increasing raids of the Luftwaffe told us the dawn of the first great attack was perilously close.

Everything was going smoothly until the Stukas appeared just as we were embarking. The familiar "Whoomph!" of the bursting bombs started Horrie growling. Urgently I ordered him keep quite. An ammunition-box being carried aboard ship was quite in order but - not with a bark inside it.

I gained the gun-deck aft and, glancing round, saw a large tarpaulin and let Horrie out of the box. The raid still thundered on and I must have looked silly sitting upon deck quietly talking to a small bulge under the canvas. But those not in the know merely thought I was "bomb happy".

When we raced safely away from harbour I let the little dog out to see upon what new adventure he had embarked; a ten thousand-ton transport, the Lossiebank, now bound for Port Said.

Next day we shared our last bully beef with Horrie. We could expect no more until we reached Port Said in about three days' time, providing we successfully ran the gauntlet of the Luftwaffe.

We had an unexpectedly good trip, and were bombed only once more. They came at us out of the sun. To the explosion of a stick of bombs, splinters whizzed across the deck, water came splashing up. The planes flew away. I noticed the little dog limping.

"He's hit," exclaimed Poppa.

A small steel splinter was embedded under the skin on his shoulder.

"It's only a Blighty," laughed Don in relief. "Horrie, you old scoundrel I believe you were only manoeuvring for a spell."

Horrie wagged his tail as if he'd accomplished something great.

We commenced the operation immediately. I pinched the skin while Don dug the steel splinter out with a knife-blade. Horrie did not whimper. He licked my hand and wagged his tail.

"We've made a good job, Horrie!" said Don and showed him the splinter.

"If only he had Horrietta here to nurse him," said Fitz.

"Perhaps he's safer as he is," remarked Poppa reminiscently.

"Tell us the romance, old war-horse," grinned Murchie.

"Go to blazes!" answered Poppa.

When we disembarked at Port Said the boys crowded round me while I carried Horrie under my arm. Once aboard the train he was allowed what freedom a crowded cattle-truck could offer. His wound was troubling him, so we took the field dressing off and he licked the wound to rights. He forgot his troubles when he heard the voices of the Arabs and immediately challenged them in no uncertain manner.

"He's got a long memory," said Gordie. "They must have given him a tough spin sometime."

From Kantara on the Canal we entrained across the Sinai Desert over which our troops had fought their weary way during the last war. Those great fighters would hardly recognize that terrible desert now. A railway runs through it, and a pipeline with water. There are stations and, in places, cultivation. Not in their wildest imaginings would the Mounted Men know that desert. Seeing what has been done with one of the most terrible deserts in the world made many of us wonder why we do not do a great deal more to develop our own interior which of course could not be compared to a true desert like this.

We travelled straight on into Palestine and camped at Dier Suneid. To an excited yelp from Horrie we saw the towering form of Big Jim and the Gogg coming to welcome us.

"He's lost quite a lot of weight during his soldiering," laughed Big Jim as he petted Horrie, "but we'll soon alter that," and he carried the pup away. He soon returned with a parcel which he opened before the ravenous Horrie.

"Officers' mess!" quoth Sergeant Poppa with a hungry glance at the good meats. Big Jim only laughed; some officer would go short of a good meal that day.

"He's grown quite a lot," said the Gogg as he opened a bottle. And we gathered round and celebrated. Big Jim and the Gogg had the Rebels camp all ready and had somehow scrounged a few bottles to do it in style.

Horrie thought he was the guest of honour, sitting in the tent listening to our reminiscences and watching the expressions on the faces he knew so well.

"So Horrie is a little Anzac!" said Big Jim in a proud voice.

Horrie stood up and marched to the tent door on guard.

"Wogs!" exclaimed the Gogg and Horrie, growling fiercely, dashed out to chase the suspected Arabs.

"He's home again," laughed Murchie.

"His belly is well lined again," grinned Fitz. "He's ready to soldier on immediately."

It was late that night when Big Jim and Sergeant Poppa retired.

"Reveille at daylights grinned Big Jim.

"Like hell!" we replied in one voice.

"I see the Rebels are back!" laughed Big Jim as he walked out into the night.

The camp settled down to routine, keeping fit while awaiting the next call of war. Daily, scores of soldiers from various battalions came to visit Horrie; many had seen him in action, while many others had heard all manner of stories about him. Horrie had become a hero dog and accepted all honours with becoming dignity, posing for dozens of cameras.

Dier Suneid district appeared interesting for the first few weeks, the countryside of yellow sand contrasting with green orange-groves hedged by cultivated cactus. The Arabs, too, were much more colourful than those we knew in Egypt, but did not appeal in the least to Horrie, who chased them on sight. They were strictly forbidden the camp, for they were notorious thieves. The soldiers had to chain their rifles to the tent poles. Despite the closest watch the prowlers would come, silent as the night itself, and work their way into the tents. They had even been known to unchain the rifles and get away with them, making not the faintest sound to awaken the sleepers. At last a lamp had to be left burning in the tents all night while one occupant of each tent must remain awake on watch. Here the Rebels were greatly envied by every other tent for we all could and did sleep the sleep of the just. No Arab, no matter how expert, could sneak within many yards of our tent, let alone right into it. The watchful Wog-dog saw to that. Horrie was no longer content to bark; he never challenged now, but simply flew at any Arab he saw and used his needle-sharp teeth.

The village of Dier Suneid lay some half-mile from camp. It was strictly out of bounds.

"We must visit it some time," declared Murchie.

From the main Tel Aviv-Gaza road the village was screened by a lovely orange-grove. It was completely enclosed by a high mud-and-straw wall, with apparently one little gate. The dark and sinister mysteries it might hold fascinated us very much. Sergeant Poppa made it all the more alluring, for he had seen hectic times here during the 1914-18 war; the wily old bird knew where our thoughts often strayed and he enlivened them by dark hints of dusky maids dancing in shimmering veils to the weird wailing of native music.

The day came when Don and Feathers and I seized an opportunity to investigate for ourselves.

"You're staying right here, my fine lad!" declared Don to the protesting Horrie as he tied him to the tent pole.

"What a lovely time he'd have in that village," laughed Feathers.

"He'd cause a commotion we mightn't care to remember," declared Don. "Right you are, he's tied securely."

We stepped out of the tent and innocently sauntered through the camp. Once in the shelter of the orange-grove, we hurried on. We emerged from the grove by the wall and through the gate could catch a glimpse of a narrow, winding street. Shrouded figures passed by now and then, a loaded camel lurched by. Feathers beckoned and we stepped through the gate. We started warily down the street, hemmed in by mud houses. Archways opened out in the inner wall which led into tiny yards in front. Arabs passed us, staring. Urchins played amongst the fowls and donkeys in the street.

"Few soldiers have been through here judging by these staring eyes, I remarked.

"A nasty place for a brawl," said Don, "to be bailed up in this rabbit warren."

"Don't forget where the gate is," warned Feathers.

But all seemed peaceful enough, although we did not quite like the crowd of Arabs now silently following. Some of the urchins now screeched at our very heels, demanding "Baksheesh, Baksheesh!"

"Well," said Feathers with a sniff, "I don't know so much about this alluring exotic perfume the dark maidens are supposed to use."

"The village does hum a bit," I agreed.

"Smells horrible enough to stop a watch," said Don.

A murderous-looking fellow stopped us as we were passing a shadowed doorway.

"Cancan?" he growled, and beckoned us to follow him.

"Nothing doing," I shrugged.

"Cognac? Very clean cognac?" he inquired, and offered a bottle of filthy-looking liquid.

"Finish money, George," I replied.

"Cognac good!" he growled. "Five hundred mils! Australia plenty money."

"Finish money, George," I answered deliberately as we walked on.

But now the urchins had withdrawn a little, cat-calls followed us, a stone whizzed by Feathers's head. We faced about, a little uneasy at the crowd now between us and the gate. Tough-looking Arabs were behind the urchins urging them on; stones began to fly.

"I'd- like a crack at that big black sod urging the Arabs on over there!" remarked Feathers.

"Looks like we'll have to fight our way out," said Don.

"Come on, better not waste a moment," I suggested.

As we stepped towards them a shower of stones greeted us; it looked ugly.

All of a sudden the crowd glanced behind then surged towards us with a yell. We thought they were charging us but heard the squawking of fowls, followed by Horrie's bark.

"Horrie to the rescue!" laughed Don. "Hooray!"

Squawks and yells answered Don as three braying donkeys and the Arabs flew past us down the narrow street, the little dog snapping at their heels. Horrie certainly had timed his blitz at the critical moment. It was staggering how those grown-up ruffians and the mob of urchins panicked on the instant; perhaps it never entered their heads that the tiny dog was not the forerunner of an angry Australian guard with fixed bayonets. As Horrie sighted us his enthusiasm became fanatical and he dashed past to complete the rout. I raced after him but he ignored me to concentrate on Arabs scrambling to squeeze into a narrow doorway. The shrieks as he snapped their bare heels could hardly have been louder had the village been put to the sword.

"Now's our time!" laughed Don as I snatched up the pup. Quickly we beat a retreat, passing two Arabs with bleeding noses; one had rushed out of a doorway as the other rushed in. But we gained the gate and leaped outside with Horrie struggling to get back to the village and into the fight again

We were very relieved, and enjoyed a good laugh all the way back to the camp.

16

WE LOSE MURCHIE

THAT night in the Rebels' tent Don casually asked Poppa:
"Ever drunk any Wog cognac?"

"Only once," answered Poppa reminiscently, "during the last stoush in these parts in '18. I only drank a bottle or two. I woke up screaming; little blue mice were gnawing me to death. A large pink elephant saved me in the nick of time, he came rushing up trumpeting fire like a flammen-werfer and trampled the mice to death."

Leave to Jerusalem was granted the battalion. Pleasurable was the excitement when the Rebels' turn came with Sergeant Poppa in charge. Very excited was the Wog-dog all done up in his new corporal's uniform. The Rebels in a body boarded the leave-bus, Horrie the first aboard.

These leave-buses each held thirty men. The drivers were Arabs and wilder men at the wheel you wouldn't find anywhere in the world. They'd simply turn on the juice, trust the lives of all to Allah, and let her go. They'd grin from ear to ear as the bus roared along with the boys yelling "Let her go, George, let her go!" They needed no encouragement; those drivers made the hair rise on my head, for a mistake of inches along portions of the hilly road meant the bus must go hurtling far below. There was barely room for two buses to crawl past one another.

The judgment of the drivers was miraculous; in truth, again and again the good Allah did look after us. A toot behind was never the signal to the driver that another bus wished to pass, it was the signal for a race. And a race it was, with the driver's grin set and his eyes sticking out like pickled onions as we took hairbreadth bends on two wheels with the rival bus roaring up beside us filled with cheering, carefree soldiers.

"Head and head!" the boys would yell as the buses drew level. "Beat him on the post, George!" "Shave his wheel, George!" "Topple him in the ditch, George!" It was impossible not to laugh and thrill even though the opposing bus was wearing us down with little more than a coat of paint separating the vehicles.

I'd breathe a sigh of relief at sight of the walls of Jerusalem far below.

Our first job in Jerusalem was to buy Corporal Horrie's identity disc. We filed into a cubicle, the metal worker squatting there tinkering at his trade as his forefathers had done for centuries. We handed him a Greek two-drachma coin, a souvenir from Greece for the purpose. The master craftsman bent over his work and with the tools of a thousand years ago neatly inscribed on the coin Horrie's name, number, and unit.

"Horrie, E.X.1. 2/1 M/G. Bn." The E.X.1 was a "special", it meant that Horrie was No. 1 warrior from Egypt. The coin, his historic identity disc with the unit colour patch were attached to the harness which he so proudly wore.

"There's never been a soldier-dog so well dressed up as he," said Sergeant Poppa admiringly.

And the Wog-Dog proudly wagged his stump.

"Better keep a tight grip on him," advised the Gogg.

We did. There was a fanatical light in Horrie's eye. Around us was the shuffling of thousands of his enemies, surly-visaged Arabs, and folk of every Eastern nation.

The sight of the little Wog-dog leading, almost pulling us through the streets of Jerusalem created an interest not unmixed with laughter at Horrie's menacing growls. From a crowd of noisy guides we selected a laughing little chap who declared he was the "son of Sandy McKenzie".

He took us first to the Wailing Wall built by King Solomon in 1000 B.C. It is one of the holiest of Jewish shrines; shrouded figures even now seemed to be clutching the base of the age-old wall. At night these old rocks are moist with a dew said to be shed by the rocks from weeping in unison with the Jews. A legend says that long ago the old wall had been lost to sight by time and the refuse tumbled upon it during the numerous battles in which the old city was finally put to the sword. In time, a shrewd king wished the wall uncovered and to save labour threw handfuls of coins upon the earth that buried it. The Jews of Jerusalem uncovered the wall by scraping the earth away while seeking the coins. Judging by the prices charged the Australian soldiers for articles they bought in Jerusalem, there is truth in this legend.

"Sandy McKenzie" next guided us to the Bazaar, showing us the old city first. We were indeed away back thousands of years ago. The bazaar was like the warrens of an under-ground city. The narrow streets were roofed after the style of an arcade into which light filtered through narrow slits in the roof. Under this medieval arrangement the light filtered down through drifting dust and smoke. Through the gloom came the shuffling of many feet as mysterious figures glided past. From the windows of tiny jewellers' shops dim lamps reflected gleams of light from jewels. Here and there, exposed to the open street, was a butcher's shop with meat swarming with flies. Fruit and vegetable stalls extended across the footpaths and even on the roads along which many little donkeys slowly came, bowed down under giant burdens. On top of these great loads solemnly squatted the owner, lord and master. Struggling on foot behind would come the wife, also loaded with goods, while behind her would trudge the children according to age, also heavily loaded. This use of women and children as beasts of burden used to make the troops very angry in Egypt and Palestine.

From many gloomy shops in the walls, dim figures bent over brass and bronze, copper and gold and silver work, squatted and smoked and waited and toiled and gossiped as their forefathers had done for thousands of years. Throughout wars and pestilence and famine, throughout plenty and desperation, throughout the uttermost heights of glory to the most terrible depths of despair has Jerusalem survived for thousands of historic years.

Black-clad Arab women peered at us over cloaks drawn up just below flashing black eyes. A hooded head-dress covered the head to the forehead; silver and copper coins hung from the forehead over the nose. Everywhere was the murmur of guttural, hoarse, soft or whispering voices haggling over prices of wares.

"I wonder what those little flags mean," nodded Fitz to small flags drooping dejectedly over gloomy dwelling holes. "Sandy McKenzie" explained.

A white flag advertised that therein hopefully waited a marriageable daughter. A blue flag meant that an unmarried mother with child waited within, ready for a personable husband who wished a family ready-made. A black flag meant a widow for whose marriage a father was willing to pay a dowry. A green flag meant that the occupant was absent on a holy pilgrimage to Mecca.

We wandered for long, intensely interesting hours in the old city; here we lived again in the East unchanged since the days of Christ.

What a contrast to pass from the old city, to the new, to walk from two thousand years ago straight into today. The new city was a place of modern buildings, and clean, wide concrete roads along which taxis smoothly hurried. The traffic was directed by tall Englishmen clothed in the blue uniform and white helmet of the Palestine Police. There were modern picture theatres. And here on the screen was flashed in English, French, Arabic and

Yiddish the interpretation of the play. Wide concrete footpaths reflected light so glaringly, that ninety per cent of the people wore sun-glasses. The modern shops displayed glittering jewellery and vivid and gorgeous silks. The sober, dark gowns of priests of many religions mingled with the bright flimsy frocks and the sporty, gay coloured shorts and closely fitting sweaters of Miss Jerusalem. Large, open cafes had their marble- and glass-topped tables and chairs extending on to the footpaths, protected from the sun by gaily coloured veranda blinds, reminiscent of Paris. The men wore grey slacks and light, open-necked silk shirts. This was a modern city of carefree atmosphere.

One night soon after our arrival in Palestine we were awakened by Horrie's frantic bark; something in that bark brought nightmare memories of Greece and Crete and instinctively we were flying for the slit trenches. The bombs whined down and exploded but nobody was hurt. It was only a lone night raider, but all the same Horrie was on the alert.

At daylight, Horrie quietly slipped out to inspect the bomb craters. He found an Arab there who was also curious. Horrie immediately attacked and returned growling to the tent with his mouth full of Arab pants.

Meanwhile, energetic training went on apace amongst a great army congregating throughout the length of Palestine. Our crowd though felt very uneasy for there were increasing rumours of a new campaign about to break out in Syria and those many thousands of us who had lost our arms in Greece and Crete were not yet re-equipped.

"What are the chances?" growled Murchie.

"Live in hopes," answered Sergeant Poppa. "You can bet your life the heads are doing all in their power to re-equip us."

But the old war-horse was plainly anxious, as were thousands of other men throughout Palestine. We did not want to be left out of any stunt.

We were only just beginning to realize fully the disasters that had overtaken our forces in Greece and Crete - the fateful losses in men, ships, and material, the miraculous fight the small Navy had put up, a Navy we now knew was fearfully battered in ceaseless fights against over-whelming odds. We realized the desperate struggle now being waged to keep us supplied with bare necessities, let alone the difficulties of manufacture and transport from Britain of great quantities of new material.

But an event occurred in the life of the Rebels that overshadowed the anxieties of war.

Horrie disappeared.

The search started in earnest. The whole battalion, or those men off duty, searched the camps for miles around armed with his photograph. Many a false clue was followed up, to no avail.

We were in despair.

On the ninth day a leave party left for Tel Aviv, forty miles away. Don Gill was with them.

They were walking down a street in Tel Aviv when an excited bark halted them. There was the Wog-dog streaking across the street, dodging the dangerously moving traffic. He leaped at Don, tried to eat him. He was in a shocking state, skinny as a rake, covered in oil and tar. The boys took him to a restaurant and fed him on the chicken in the sandwiches. They ate the bread themselves.

We never learned how Horrie had travelled that forty miles to Tel Aviv. He may have been commandeered by some foul renegade; being a Rebel dog he more probably went A.W.L.; simply jumped on a leave-bus, and took a trip to town.

Tel Aviv is a modern city, although thirty years ago it was but desolate sand dunes. The Jewish inhabitants of Jaffa built the city from money supplied by the Jewish National Fund. The Jews mostly moved to Tel Aviv, leaving ancient Jaffa to the Arabs.

In our wanderings, we strolled over many of the last war battlefields of Palestine. One of these was Gaza, city of Samson.

Numbers of the stone houses still bore signs of our shelling while on the slopes of Ali Muntar we could still find souvenirs of the terrible fighting that took place on that grim Turkish redoubt. Throughout both Syria and Palestine, as we travelled by train and truck, we used to wonder again and again how the Anzacs successfully fought for hundreds of brazen miles right through this country. They had only horses for transport. We had trucks and tanks and trains and aeroplanes. Their only water in the Sinai Desert was in the far scattered oases, while we had reservoirs and a pipeline beside us. We often wondered how they did it and pushed a fighting enemy back ahead of them.

But near tragedy was looming over the Rebels after all this time of close comradeship. Action was soon to break out across Syria and the Rebels would not be in it.

Many of the troops were not yet re-equipped. We had to remain in Palestine. Murchie had become more and more restless, Poppa more and more gloomy. Sergeant Poppa had an idea that no war could be properly won unless the Rebels were in the vanguard.

"The 2/3 M/G. Battalion looks like being in it," said Murchie. "What about applying for a transfer to their unit and seeing the fun?"

"Easier said than done," answered Poppa. "There are enough boys applying for transfer to stock a new army."

"How about Horrie?" I asked. "He may not be accepted as part of their show like he is here."

"We couldn't go anywhere without him now," smiled Don.

Horrie sat on my bed with his ear cocked, listening intently.

"The little fellow understands every word we say," said Poppa as he patted the Wog-dog.

Horrie's expression certainly meant, "Don't leave me behind, please."

We talked until long into the night, weighing the pros and cons. But in the long run it was only Murchie who succeeded in gaining a transfer.

"After the show in Syria is over I'll rejoin you blokes quick and lively," he promised cheerfully. We knew he would if possible and we were glad to see him his old cheerful self again. Murchie was born for action. But we were awfully sorry to lose him and he tried to hide his sorrow at leaving us, even temporarily. His last words were: "Good luck, you chaps; and good luck, little Wog-dog."

Horrie knew he was losing a staunch friend; he stood on his hind legs and rested his paws on Murchie's knee as Murchie knelt to pat him. Long after he was out of sight Horrie sat outside the tent gazing in the direction Murchie had taken.

We never saw Murchie again. In the far-flung movements of a world war, units once separated do not always rejoin.

Murchie is in Java now. We picture him as still holding out in the hills with some wild band of guerrillas, fighting jungle ambushes as he fought with his battered truckload of New Zealanders and Greeks during that desperate retreat from Greece.

Somehow we feel confident that we will see Murchie again; after it is all over, the Rebels will stage their grand reunion.

17

HORRIE FALLS IN LOVE

AFTER the Syrian campaign our unit was moved to a camp at Khassa and here was born Horrie's great romance. We camped alongside the 2/1 Anti-tank Regiment, just returned from the campaign. I strolled across to locate Bruce McKellary, Horrie trotting beside me, when some one yelled "Imshi!" I took no notice, as this was the command to some stray Arab to "Clear out! - get to hell out of it!"

"Imshi!" yelled that abusive voice, as if at me.

"What's biting you?" I called to the soldier.

"It's O.K., Dig," he laughed, "don't get your wool off."

"What's the matter?" I asked as Horrie and I strolled across to him.

"1 thought that was our little dog Imshi," he replied. "He's the dead spit of her - with reservations."

"I knew there was a lady dog of his breed in Crete," I said. "Surely there's not another in Palestine."

"Wait till you see her," he replied. "You wouldn't know them apart, except that she's got a longer tail. She's our mascot. Been with us through Egypt, Greece and Syria."

"Horrie's been with us in Egypt, Greece, and Crete," I replied. "Been sunk on a transport too."

"That evens up the record," he laughed.

"Horrie's got a wound stripe," I added.

"That goes one better," he said glumly, "although I'll bet your Horrie hasn't wet his pants as often as our Imshi has!"

"Horrie's been in plenty of ticklish engagements I replied, "and if he couldn't find a tree he used my boot."

I located Mac and we enjoyed a great yarn. Noticing that Horrie had disappeared I asked, "What's this Imshi of yours?"

"The gamest little dog in the world," he laughed, "hardly the size of a woolly elephant but with the guts of a lion. Why?"

"I've heard about her."

"The whole Army has," he said proudly. "You'll see her later; her valet is the cook, and he puts more science into preparing a meal for her than he does for the troops."

"She probably appreciates his efforts more," I laughed.

"Oh, he's not a bad cook," said Mac.

And then Horrie appeared with Imshi. With the most innocent expression, tongue gently hanging out, one ear at the cock, one at the flop, he trotted up to introduce Imshi to me.

She was following behind.

"Struth!" declared Mac, "he's pirated her!"

Most obviously so.

Oh, the he-male of the species! No longer was he a sea-sick hero playing second fiddle to Ben the Seadog. Here, he was Horrie the Wog-dog on solid ground with the bashful Imshi following behind.

I stooped down and smiled and patted Imshi; she responded bashfully, a "come-hither" look in her large brown eyes. I laughed, having been caught before.

Horrie fussed about her, his every attitude informing her I was quite all right - "just one of the boys."

"I dunno," said Mac thoughtfully, "but this love at first sight may cause trouble."

"We are in the land of Antony and Cleopatra," I said. "They both passed this way."

"Yes," replied Mac, "and what was the end of the journey?"

But Horrie, fussing round and round Imshi, had no misgivings. As to the winsome Imshi, she cared not a rap.

Imshi, though trustfully timid, was game too; she was of the same breed as Horrie, and an attractive wench indeed to a Wog-dog even though a shade taller, and of the same colouring and build, with her proudly carried tail obviously longer.

Having captured me she turned all her wiles on Horrie, showing off her paces while pretending she took far more interest in me than in him.

"She's bent on conquest," said Mac resignedly. "He's a goner."

"It always works out that way," I answered, "when the lady has made up her mind."

Imshi now blatantly set herself to capture the Wog-dog's heart. Having further won my admiration she pretended not to notice the admiring Horrie but pranced away around us with head playfully erect, fox-trotted back to us, and then with a sparkle in her eye bounded away behind a tent.

Ah! but if Horrie did not trot after her she would immediately reappear.

When Horrie was not chasing this provocative miss he sat at my feet voicelessly admiring her.

"Go your hardest, Horrie," I advised. "She is only kidding you!" and Horrie would trot after her, catch her up and sit beside her while both intently gazed at the horizon.

"Just a bit of finesse!" grinned Mac.

Horrie, anxious at getting nowhere, sniffed inquiringly at her face. she disdainfully turned her cheek.

Mac chuckled.

Horrie playfully took hold of her ear but Imshi scornfully turned her head aside. Horrie playfully tugged her ear but she snapped at him and he bashfully shuffled back to squat and view her inquiringly, little head to one side.

"No ordinary street pick-up, this lass," I grinned at Mac.

He grinned back with a wink.

"I bet she lands him!" he said confidently.

She did.

Horrie, puzzled by Imshi's attitude, trotted back to me with comical look of inquiry. Imshi gave a sudden yelp and raced after some mystical object in the distance. Horrie wheeled around and excitedly raced after her.

"He's a goner!" said Mac with finality.

The two little white figures gradually disappeared across the sandy waste.

"That's done it!" said Mac in triumph.

I returned to camp.

Very late that night I felt Horrie quietly climbing up to the foot of my bunk.

"Imshi's a nice little dog," I said, and felt his stub tail wagging in agreement.

The two became inseparable. If Horrie was missing I'd find him with Imshi away across at the Anti-tank regiment. If Imshi was missing from her rightful camp, she'd be playing with Horrie at our camp. It was sweet romance but the practical Horrie combined belly love with it; he invited Imshi to come across and share his bone. she accepted somewhat disdainfully. Then she invited him across to her camp to taste real meat. Horrie accepted with an alacrity that hinted of planned purpose beforehand.

"He's a little strategist," laughed Poppa. "He angled for that! He knows Imshi's cook boss can give him much more tasty meat than we can."

"And plenty of it," added Gordie. "Horrie will be a regular visitor at the Anti-tank cookhouse."

And he was.

As the days passed Horrie found himself torn between his desire to accompany us on our route marches and his longing to dally with Imshi. He tried his hardest to coax Imshi to accompany us but she was adamant; active service experience had evidently taught her that camp was the safest place and she was determined to stick to it. Away we would march, with Horrie gazing longingly at the column and then running back to Imshi in a last effort to persuade her to come along. He would coax all he knew until we were a long way off; then, with a resigned "Well, will be seeing you tonight!" he would come scampering after us as fast as his funny little legs could carry him. At the head of the column he would take his place beside Big Jim and gaze up with little tongue hanging out, and trusting brown eyes full of wisdom as he waddled his quaint little body into step.

"So you've decided to join the parade!" Big Jim would growl down at the tiny dog. And Horrie would gaze up with a wag of his tail and knowing wink that plainly said "After all, soldiering must come before pleasure". Then he'd look back at the grinning Rebels, his tiny feet and comical waddling body keeping time to the tramp, tramp, tramp, tramp .

"You're a dinkum Rebel all right," Big Jim would scold. "'Shun! Eyes front!" and the Wog-dog would pay attention to the job ahead.

But we had to put our foot down on Horrie's midnight romances, on the score of pressing military necessity. To enjoy a full issue of sleep and at the same time save our rifles from Arab thieves we now found it necessary to chain Horrie up every evening. Otherwise the night would call and he would throw duty to the winds and vanish to the desert with Imshi.

Antony and Cleopatra caught the moon madness upon these very sands. Our hearts were all with the little Wog-dog and Imshi but - we had to safeguard those rifles.

Imshi, once assured it was now impossible for Horrie to escape at night, very sensibly remained in her comfortable quarters near the cook in the Anti-tank regiment. But first thing in the morning she would call for him, we would let him loose, and away she would take him to her friend the cook for breakfast. When they could eat no more, away they would trot on a lizard-hunting expedition for the day. At least I hope so.

Time drifted on with all troops engaged in strenuous training. Then - into the idyll of the Wog-dog came the Big Bad Wolf. A huge, gaunt, savage Alsatian.

He came prowling into the lines of the Light Aid Detachment, a famished outcast. No man would turn a homeless, hungry dog away, at least not until he had given him a feed. And the Alsatian was so glad of a meal that he simply refused to leave. So the L.A.D. adopted him. He immediately repaid it by becoming an excellent watchdog; no prowler dare enter the L.A.D. lines even on the darkest night, he would have been torn to pieces.

As this was the Khassa area, the L.A.D. named their huge wolf-dog "Khassa". The name fitted him to a T., his giant, gaunt frame, his snarling teeth, the glare in his eye all fitted in to "Khassa the Goliath". Worthily he guarded the camp of his benefactors, the L.A.D.

However, those insignificant insects, the Wog-dogs Horrie and Imshi, were in opposite camps, hence they and Khassa were enemies in Dogland. And the cheeky Horrie made no bones about his dislike for the new arrival.

"I hope that little wretch doesn't buy a fight with Khassa," remarked Fitz one day.

"If he does there'll only be one mouthful in it," said the Gogg grimly, "and that mouthful will be Horrie."

I was very uneasy, as were all the Rebels, for the little Wog-dog's safety. He had long since grown the idea that he was an army on his own.

"If Khassa should take a fancy to Imshi," laughed the irrepressible Feathers, "there'll be skin and hair flying."

Even as we were talking Khassa on giant legs came boldly towards our tent. Horrie leapt up with hackles erect, growling menacingly In no uncertain manner he warned Khassa keep clear of the tent. Khassa stood and glared down at him.

"He's wondering whether the Wog-dog is worth just one bite," said Poppa. "Don't do anything, perhaps they may be content with an armed neutrality."

But they were determined on fight and war was declared a week later. It came about through a lady.

Imshi was trotting across to visit Horrie when she was waylaid by Khassa. Horrie was dozing when Imshi's yelp for help urged him out the tent. He saw his girl friend in distress and charged straight at Khassa who wheeled from the cowering Imshi to face the insignificant hero. The big dog stumbled and Horrie gripped his hind paw and used war service and tactics to escape annihilation. As the big dog yelped and snarled, Horrie hung on, dragging the leg round. The long body followed round and round, the Alsatian's fangs striving to reach the tiny tormentor. Horrie clung on like a vicious rat to a lion's hind paw; try as Khassa would, his snapping fangs could not reach him. Imshi had backed away to watch the gallant Horrie, but now she flew to the rescue and sank her teeth in the neck of the big dog. As his neck straightened out in fury, she darted away. The instant his head snapped back towards Horrie that instant she flew back at his neck. It was perfect team-work and in the first swift scurry of attack the Goliath was mercilessly outwitted. But in a matter of moments the powerful Khassa would turn the tables and here Frank Bruce came running to the rescue with shouts and blows trying to break the mix-up. Frank separated them but received a nasty gash in the arm from the infuriated Khassa. The big dog limped away snarling with Horrie standing protectingly in front of Imshi, growling to Khassa to "Keep moving!" When the danger had faded away Horrie received a pat from Frank in recognition of his bravery; then with his tail mast-high and hackles erect, he proudly escorted Imshi back to the Anti-tank lines. Poor Frank, nursing a very painful arm made his way across to the R.A.P. to get the wound stitched.

But life is a whirligig of hopes and sorrows, of joys and setbacks, and even lovers must part. The Anti-tank regiment was ordered to Syria and Imshi marched with it.

A woebegone lover, Horrie haunted the abandoned camp site. As the days passed and she did not return he even refused his tucker, and grew thin, scrawny and miserable. Then he took to roaming the countryside, visiting the scattered camps and searching for his lost love.

"We'll have to do something with that young Romeo," declared Poppa, "or he'll fade away into the desert dust."

"What did you do under similar circumstances!" grinned Feathers.

"Took it like a man!" declared Poppa.

"A fine mess it's made of you!" remarked the Gogg.

"I could point out some men," answered Poppa impressively, "and, mind you, I needn't go past the Rebels, who are such a mess that even years of soldiering will never make anything out of them!"

"How many years of soldiering?" asked Fitz innocently.

But Horrie disappeared.

We searched for him far and wide.

"Women always mean trouble!" sighed Poppa wearily, after a long, exhausting search.

"If he's followed the Anti-tank regiment, he's got a long way to go," said Feathers.

"I wish the little beggar would turn up," sighed Don.

He did, on the evening of the fifth day. Some of the boys returning from leave had picked him up on the road to Jerusalem, fifty miles away.

Horrie was in a pitiable state, skinny as a rake, footsore and weary. He tried to kiss each of us at once. He ate like a wolf and lay about and rested a few days. Soon he was on the way to becoming his old self again.

"How soon they forget Love!" sighed Poppa.

"You're forgotten long ago," grinned Fitz.

"That's something, anyway," answered Poppa dreamily; "so far as you nit-wits are concerned I don't think you've even been remembered."

But Horrie had by no means forgotten Imshi. Whenever her name was mentioned he would jump up with brightening eyes, tremblingly excited.

18

THE HAZARDS OF PEACE

DURING November big news excited us - we soon would move to Syria. From returning troops we heard tales of icy cold.

"Horrie is a desert dog despite his Greek campaign," said Feathers. "How will he stand up against bitter winds and snow and sleet?"

"Make him a uniform," ordered Poppa, "for service in Arctic regions."

Don offered an old greatcoat he had scrounged from a salvage dump. We cut off the tail, stood Horrie on my bunk and draped the piece of coat over him. He appeared dubious but guessed something new was in the wind.

As I snipped off the material at his stern end the Gogg advised, "Be careful of his rudder."

Horrie instinctively sat on his tail, so protecting that appendage from the scissors.

As we got time off duty and the uniform began to take shape we grew quite keen, as did the patient Horrie. We fitted the head and body and legs really neatly, allowing for a row of buttons down the belly. But then came a problem.

"How about his fly?" inquired Fitz, "he can't undo buttons, but he must have a fly."

"Could we make it like the sailors' trousers?" asked Feathers doubtfully. "Sailors appear to manage without buttons. I don't know how they do it, myself."

"How about Scots style - kilts" grinned the Gogg.

"Nit-wit!" snapped Poppa. "Scotsmen wear pants."

"How do you know?" inquired the Gogg.

"Anyway, the Wog-dog must wear pants," said Don, "otherwise he will freeze."

"I don't know what we are going to do about the tail, either," said Gordie. "It's only a stub of a thing and all gristle, but he uses it quite a lot."

"He uses everything quite a lot," said Fitz.

"To wrap his tail up might spoil his carriage," insisted Gordie.

"I don't know about the carriage," grinned Fitz, "but it would spoil his wagging.

"Funny one," growled Poppa, "you ought to be in the transport section."

"Wish I was. I'd dodge your tongue-waggin'," replied Fitz.

"Get on with the job," growled Poppa.

We overcame the difficulty by cutting a half-moon in the cloth around Horrie's rudder. Similarly with the other little difficulty. To the top of the neck-piece we sewed the unit colour-patch, braided the edges with tape, and made all complete except for the buttons by sewing two little gold corporal's stripes to the tunic over the right leg. We'd been wondering how we could obtain the buttons but a brainwave solved it. I engaged Feathers's attention by pointing out the excellent fit of the uniform while Don quietly snipped the buttons off Feathers's tunic which was hanging up on the tent pole. Feathers always did keep his buttons clean and shining.

Horrie was quaintly pleased with his new uniform; proudly he stepped out to do the rounds of the camp and show it to his many friends. They admired him so much that he swelled out until the suit nearly lost its buttons. He returned to us fairly treading on air but immediately growled menacingly and began searching to locate an Arab, for an Arab voice was harshly resounding throughout the tent. Then he made a dash for the wireless and I had to switch off the station.

It was about this time I received a mouth-organ from home and proceeded to enliven the Rebels' monotony.

"Can you play 'Far, Far Away?'" groaned Don.

"I'll try," I replied hopefully.

"Well, there's the direction!" and he pointed out the tent to the desert.

But Horrie was appreciative. His was a musical soul and he loved to sit back and howl accompaniment as only a Wog-dog can.

"You should train Horrie to sit up and hold your hat in his mouth," suggested Fitz. "It will earn you many a feed after the war."

"It's about all any of you will be good for," sighed Poppa.

"I can see you grinding a monkey-organ in Pitt Street," grinned Fitz.

"It's you will be holding the hat in that case," replied Poppa.

"I'll paint you a sign," volunteered the Gogg to me. "'Spare a coin for an old digger.' You can hang it round your neck while Horrie collects the coins."

But Horrie was faithful to me and my music.

At this period of our training, leave was fairly plentiful but on a percentage basis. This meant that seldom could more than two of the Rebels go together. When they returned, two more would go. Horrie always went. At the mere mention of leave he became very excited.

"Yes, Horrie, you can go if you know where your uniform is." And Horrie would run to the tent post and try to climb it in his eagerness to put on that uniform. Horrie was far and away the most travelled dog in Palestine; he had seen more of it than the majority of the troops. Its principal cities and many of the towns and villages were familiar to him. But now when he strutted about in his fine new uniform he became better known than ever. Troops that we seldom were in contact with recognized him, and civilians too.

"That's the little Wog-dog!" was a common exclamation as Horrie, at the head of a leave party, would march proudly along a street. Camps, soldiers' clubs, popular cafes, beaches, dance halls - wherever soldiers congregated Horrie was a familiar sight. If the boys turned into any of the numerous wine and spirit cafes in any town Horrie was there too; he'd sit up on the counter and drink his glass of milk with the best. No matter how many "shouts" there were Horrie would always accept his glass. Never would he let it be said he couldn't take it. He would waddle out of the cafe blown out like a barrage balloon.

But he was always the soldiers' dog; he would take no familiarity from any one except an Australian soldier in uniform. The Jewish and Arab cafe owners often tried to pat him but received in thanks a warning growl, as did civilians in the street. One day Don and I escorted two Australian nursing sisters on a tour of old Jaffa and although we spent the entire day with them Horrie most ungraciously would not allow them to pat him. He was a dyed-in-the-wool Diggers' dog and he recognized the Digger uniform only.

It was when due for a trip to Jaffa that Poppa advised me, "Be sure and visit Elijah's cave."

"Why?"

"Because the Jews reckon that any one unbalanced in mind becomes sane again if he visits the cave."

"Thanks. Isn't it time you paid the cave a visit?"

"I did," grinned Poppa, "'way back in 1918."

"Doesn't seem to have done you much good," remarked Fitz.

"That's because you can't distinguish the difference," replied Poppa. "You're overdue for a visit. In fact, it's so noticeable that the O.C. has ordered me to take all the Rebels there."

I did not visit Elijah's cave, but the Jews also say that any one suffering from nervous disorders becomes cured after a few days there. The cave is at the foot of Mount Carmel. The Arabs call Elijah the Prophet Al-Hader which means "The Green One". He was popular, and will be forever green in the memory of the Arabs.

A somewhat embarrassing incident occurred when I was booking in at the hotel at Haifa. The proprietor who spoke imperfect English inquired if I wanted a room "with or without". This could mean various things; perhaps it meant with or without a bathroom, or telephone, or anything. I thought he must have meant with or without the Wog-dog. So I answered "With".

Later, I'd comfortably snuggled down in my virtuous couch when the door softly opened. A vision glided in and paused with a hesitant smile at the menacing growl from Horrie. In the half-light from the passage way her big, dark eyes smiled inquiringly at me. Noiselessly the door closed and she was there. For a long startled moment, I did not realize that this was the "With".

After some wretchedly embarrassing explanations in which I pictured myself torn to death by a sweet young thing scorned, I managed to get rid of her with a suitable present. To make security doubly sure, I 'phoned the proprietor and impressed upon him that I had changed my mind, I wanted the room "Without".

The puzzled, silly ass again mistook me. I had just dozed into uneasy slumber when a growl from Horrie warned that trouble had come again. I opened startled eyes to find the room commandeered by a complete harem. You can only realize such a situation by experiencing it. Smiling beauties glided towards me, drifting all around the bed, giggling softly. Several of them became embarrassingly insistent.

In an awful pickle I tried to explain I preferred to sleep with my dog for company but they burst into silvery laughter and politely insisted they must have misunderstood me. Understandably so, for there was no comparison between their charms and the angry rumblings of the Wog-dog. The more I stuttered the more they chattered and giggled, trying to understand just what variety of girl I really did desire, and arguing around the bed while I huddled there in a perfect funk, hanging on to the indignant Horrie. Then they commenced giggling and chuckling and pointing at the outraged Wog-dog until neither he nor I could stand it any longer. I simply cleared out. It was the quickest evacuation of my military career.

At the next hotel I most definitely impressed upon a sleepy proprietor: "Without!!!"

I found that the "With" or "Without" is the custom even at the best hotels in Haifa.

But you may be sure I never mentioned that evacuation to the Rebels.

I overstayed leave as usual, and to make up time decided to risk the Wog bus. This bus was always full of Arabs returning from Haifa, but an uncomfortable ride in it meant I would save a few hours. The bus was loaded with a choice assortment of Arab cut-throats. Horrie was hotly indignant at being forced to mix with such company and expressed his sentiments in no uncertain manner. His growls and teeth-showing were acknowledged by dark scowls at us both. Those Arabs would have loved to draw a knife very slowly across Horrie's throat. Keeping a tight grip on the little fury, I manoeuvred for a seat at the back of the bus so that should trouble develop I would have only one front to protect. To cop what could easily develop into a nasty situation, a Digger very much the worse for wear staggered in just before we pulled out from Jaffa. Though Jaffa was then very much out of bounds for troops, it was easy to see the Digger had been hitting the high spots. Giving scowl for scowl he shouldered his way up through the bus and commandeered a seat in the middle. He hadn't noticed me away at the back but obviously he didn't care if he was alone amongst all the Arabs in Palestine. The bus started off and my anxiety increased as the Digger commenced to abuse the Wogs who retaliated with Arab curses. I shouted warnings to him, being anxious to keep the peace until the bus reached the main road, where help might be handy in the event of trouble. But an Arab spat in the Digger's eye.

He promptly punched the Wog and the bus was a howling bedlam of struggling humanity enlivened by the barking of the Wog-dog and roars from a Digger being kicked and clawed and trampled to death. I grabbed a fire extinguisher to swing it as a club but it started to spray and filled the bus with fumes. I used it like a Tommy gun and the Wogs panicked under the fumes; in one mass they surged to the front of the bus with Horrie at their heels. During this excitement, there was an awful swerve and crunch as the mob upset the driver and the bus thundered off the road. I went head over heels but still sprayed all around me, nearly suffocating Horrie who, with a demented Arab, fell across me. Fortunately the bus did not overturn but came to a bumping stop with gasping, nearly suffocated Arabs leaping off and making for the horizon. With a screech of brakes several big Australian trucks had pulled up and the men leapt down into the melee and grabbed the last of the nearly blinded Arabs. I yelled to the Aussies to lay off, for it really was all the Digger's fault. Gasping for my own breath I was still hazily wondering at the miraculous escape. We slung the digger into a truck and pushed off before the Red Caps should arrive. The Digger was unconscious; he'd received enough blows on the head to have killed three men but the luck of the drunk stuck to him. The boys in the truck dropped me near camp and took the Digger to a First Aid Post where I sincerely hoped he'd be forced to swallow the stiffest dose of salts ever administered.

19

IMSHI TO THE RESCUE

JUST before we moved off for Syria, Big Jim walked into the tent.

"Could you have Corporal Horrie here on parade at 2 p.m.?" he smiled. "An unexpected visitor wishes to see him."

"Who is he?" I asked.

"You'll find out at 2 o'clock. You'll be surprised," and Big Jim walked away.

The surprise was Captain J. J. Hindmarsh. We thought him a prisoner of war with the Germans in Greece, but by some mysterious channel he had escaped and had now caught up with the unit in Palestine. Horrie rushed him with delighted tail-wagging.

"Good little Wog-dog," said the captain, "So you remember me." Horrie certainly did.

"I've often wondered about the little Wog-dog and whether you'd managed to stick to him," smiled the captain after our greetings were over. We were doubly pleased; the captain had always been popular, as was our present O.C., Captain S. Plummer. We had been considerably disturbed at the news that Captain Plummer was to be transferred to a new command, not only because he was a popular O.C. but also because he was a friend of the Wog-dog. We had been very uneasy as to whether the new O.C. would have any time for Horrie. But the problem was solved. Captain Hindmarsh was to be O.C. and Horrie's luck was holding.

Soon, the long convoys were moving through Palestine into Syria, Horrie and ourselves well pleased to be on the road again. At Damascus a beautiful avenue of gum-trees drew delighted calls from us. The convoy stopped a few moments and every man breathed in the clean, sweet scent of the gums. What lingering memories of home they brought.

Damascus is said to be the oldest city in the world. Whether or no, it is mentioned in the Book of Genesis as existing in the time of Abraham. It to us was a very colourful city, the effect being added to by all nationalities in their national costumes. Persians, Moslems, Afghans, Arabs, Kurds, Turks, Jews, Jebel Druses, French - all manner of peoples and customs and languages. Each year thousands of the devout pass through here on the long pilgrimage to Mecca. One name for this city of thousands of historical years is the "Gate of Mecca". Another is "Pearl of the East", and "City of Many Pillars".

We travelled north along the Lebanon Valley to the town of Baalbek, in faraway days the city of Heliopolitan Zeus, a city of pagan worship when the great temple was destroyed by lightning and fire. The Romans took the city, many nations in turn fought for it and captured it, and three big temples which formed the Acropolis of Baalbek were turned into a fortress by the Moslems. If we could but turn back the pages of time, what fascinating stories of mankind we would see.

Towering above the old city of Baalbek still stand six huge columns of the Temple of Jupiter. I wonder how many millions of vanished people have knelt there.

As we travelled up the Valley of Lebanon, Horrie lost interest in the passing scenes. It was bitterly cold and the mountain-tops were white with snow. At the small village of Zaboude we camped in Nissen huts for seven weeks; the snow slid off the rounded roofs which were anchored by cables to prevent the hurricane winds from blowing them away. In spite of his warm uniform, Horrie felt the cold keenly, and to get warmth, thought out a plan of his own. one morning after stand-to, he came shivering out of the hut with a question mark in his eye and a sock in his mouth. Trotting up to a soldier he scampered away to return and plainly say, "Chase me!" The soldier did so and we laughed at Horrie's elusive dodging. Soon, we were all chasing him. He played these "chasings" each morning afterwards; he was ridiculously small and we never could grab him. It was always we who became tired first.

We liked the natives of Syria. They appeared to be a refreshingly decent type. But Horrie detested them and those who visited us from the village avoided him like the plague. They were sports and eager gamblers. We were surprised at their aptitude for our Great National Game until they laughingly reminded us that Australians had been here before - in 1918. Those Aussies left some good Aussie cuss words and the Great National Game, besides other things scattered throughout Sinai, Palestine and Syria. These locals eagerly joined in any game of two-up, gambling heavily but shrewdly. The boys called them "grouter betters" because they would wait until three or four heads or tails fell in succession before placing their bets. They thus put their faith in the law of averages that the next fall of the pennies would probably turn up the opposite to that which had fallen in succession. They were very lucky too.

Another habit the old-time Diggers had planted here was the "Toss". Often when troops or natives could not agree upon the price of some souvenir, the native would suggest "Toss?" It was always agreed upon, and they would put the disputed price to the toss of a coin.

But rain and mud came, and an icy wind howled up the valley. Then something happened which plunged the Rebels into gloom, and spread anxiety throughout the entire camp.

Horrie became sick.

We did everything we possibly could for him. We wrapped him in our warmest blankets, kept rocks heated in a fire and put them under his bed and replaced them with fresh warm ones when they grew cold, tempted him with a special diet of milk and porridge, and gave him all the comfort and petting we possibly could. But he was very, very sick.

"Something will have to be done about that blasted door!" snapped Poppa. For despite all his warmth, when the door opened even for a moment Horrie shivered in the icy draught. The hut was heated by a small wooden heater but there were so many inquiries about Horrie's progress that the door was continually opening.

"Shut that b--y door!" roared Poppa again and again as some luckless caller poked his nose in to inquire about Horrie.

"We'll print a two-hourly bulletin," declared Poppa, "then they needn't open the door. Here Gogg, print a damn big notice DON'T OPEN THIS DOOR! Paint it on a board and hang it up outside the door; then we'll nail a two-hourly bulletin underneath. After that, if any nit-wit dares open that door I'll shoot him!"

We fixed it that way. On the board under the notice we tacked the bulletin twice daily, sometimes oftener. The bulletin was written on an envelope. One I have beside me reads thus: "This door when opened admits a cold wind which lessens our chance of keeping the Wog-dog warm. The condition of Horrie continues grave. He answers to his name only by opening his eyes. He accepted only a little warm porridge during the morning. Everything possible is being done to keep him warm, so don't muck it up by opening the door. Further bulletin will be issued at 1700 hours (5 p.m.)."

But we grew increasingly miserable over Horrie; he was a very sick little dog. We hardly spoke above a whisper and stayed awake at night keeping the stones hot for his bed. And then -

we dare not let ourselves think he was dying. We were near the depths of despair when the miracle happened.

Imshi arrived. A very lively little Imshi in a beautiful new uniform, snug and warm. It happened this way.

I was on duty at our signal office, a little tin shed, when in walked Bruce McKellar.

"Where on earth did you spring from?" I cried.

"We just arrived," he grinned, "the complete 1st Australian Anti-tank Regiment. We've to camp quite near you. How's the Rebels and the Wog-dog?"

"He's very ill," I answered. "Seems to have caught a severe chill and we don't seem able to do anything about it; we're very worried. By the way, is Imshi with you?"

"The section she's with should be here any time now."

"By Jove, that's good! I wonder if the sight of her would brighten Horrie up a bit!"

"I'll bring her across as soon as she arrives," promised Mac.

At midday, with Imshi under my arm, we strode across to the hut. The Rebels were all there with some signaller cobbers clustered around the heater. There wasn't a whimper from the tiny figure wrapped up in the blanket by -the fire.

"Horrie," I said cheerily, "look who's here!" and I put Imshi beside him.

She sniffed inquiringly at the swathed up little head, then yapped delightedly and energetically kissed his nose.

Horrie's eyes opened, opened wider. Then he struggled up, she kissed his nose; he wriggled out of the blanket. His tail waggled furiously, he kissed Imshi, she kissed him, Horrie yelped in delight and pranced around Imshi and kissed her again.

"You damned little malingerer!" exploded Gordie.

We gazed at the little corpse so suddenly come to life again; it fairly took our breath away.

"Well," said Sergeant Poppa slowly, "I've seen a good many wars, a good many years soldiering, I've seen a good many dinkum malingerers but this fair takes the bun!"

"Love's magic touch!" chuckled Fitz as the two excited dogs kissed again.

"Love me eye!" growled Poppa. "I'd just love to give him pack drill and fatigues!"

Neither Horrie nor Imshi took the slightest notice; they were busy now examining each other's uniforms with a critical approval comical to watch. Imshi admired the braid and the corporal's gold stripes upon Horrie, who strutted his stuff wonderfully.

"Here," I said, "bed is the best place for sick little Wog-dogs!" and seized him and tucked him securely back in the box.

"And what's more," declared Don, "stay there and don't attempt to get out!"

"That'll teach you to put the old soldier act over us!" grinned Feathers.

Horrie popped his chin over the box and gazed appealingly at us, and Imshi.

"Take your little flirt back to Camp, Mac," I said, "in case Horrie's pulse rises a bit too much."

That evening a special bulletin was issued on the notice board: "The Wog-dog's girl friend, Imshi, is now at his bedside. Don't panic. The services of the padre will not be needed. Horrie now shows a decided interest in life after this visit of his girl friend. Wouldn't you?"

I was first up next morning and on opening the door there was Imshi waiting to visit Horrie. She trotted across to his box and he gazed at me pleadingly.

"Oh all right," I said, "off you go, you little rascal," and Horrie leaped out of the box and away they trotted to investigate the bushes round the camp.

Horrie's recovery was miraculous; we certainly had been fearful we were going to lose him. However, Imshi worked the miracle.

Several days later, in the early morning, Horrie's familiar bark sounded at the door and one of the signallers opened it for him. In he trotted with Imshi at his heels both covered from head to tail in red mud and looking very pleased with themselves.

"You dirty little wretches," exclaimed Don as they joined us at the heater. "Your uniforms are sopping wet too."

"Better take 'em off and dry 'em," growled Poppa, "or we'll have both of them sick in bed next. One at a time is bad enough."

"If we put 'em in together," grinned Fitz, "they'd never get up."

"You dirty little grub," scolded Feathers as he scraped mud off Horrie's uniform, "why can't you be a dinkum soldier, all spick and span?"

But Horrie only grimaced and took it as a compliment, as did Imshi while Don tried to scrape the mud off her. Just then her master's voice floated across to us from the anti-tank camp.

"There you are, now you're in for it!" said Don, as Imshi obediently trotted to the door. I opened it for her while Feathers held Horrie.

"You just stay here until you're dry," he scolded. "No more scaring the life out of us with a bad cold."

"What about her coat!" asked Don as Imshi vanished.

"Let Horrie take it to her," suggested the Gogg and taking Imshi's uniform down from the drying-rack, he handed it to Horrie. "Take Imshi's coat to her," he ordered, and Horrie eagerly took it in his mouth and raced out the door.

"Hold on a minute - here's Imshi's coat!" Don yelled to Imshi's master.

Imshi and her boss halted and turned round; Imshi trotted back to meet Horrie. But instead of dropping the coat, Horrie invited her to play.

"Hand that uniform back, you scoundrel!" we heard Imshi's master shout. But Horrie held the uniform invitingly and as Imshi reached for it he ran back tantalizingly. Imshi followed and he ran back a little way again. We laughed from the doorway as Imshi's master chased Horrie, who made back for the doorway with Imshi at his heels. They made straight for the heater where Horrie sat down. Imshi gazed at him a moment then sat down beside him.

"The cunning little blighter has brought her back again," laughed Gordie. When her master came puffing into the hut we explained Horrie's scheme.

"You don't have to tell me anything about it," he declared. "The little so-and-so spends half his time kidding her away from our camp; hes cunning as a bagful of monkeys."

"Imshi doesn't seem to mind," said Fitz.

"I know she doesn't, and I've warned her about his ways a score of times."

Grim and forbidding, the mountains rose sheer up from the valley to snow-covered tops. They were rocky and barren, with scarcely a tree, but covered in places with scraggy scrub. Sometimes through the icy night the creepy howl of a wolf would come floating down from the mountains. We'd often wished for an exploring trip amongst those rocky fastnesses with maybe a shot at wolf, cheetah or gazelle. So one morning very early Don and I set out. Immediately we were up against the problem of Horrie, joyously leading the way.

"What on earth are we going to do with him?" said Don.

"Blessed if I know," I replied. "He wouldn't make much of a mouthful for a wolf."

But we'd started, and hadn't the heart to order the little fellow back.

He was on the hunt all right and immediately nearly got us into trouble by rounding up a score of huge blacks. Previously we had taken considerable interest in these cheery giants on the few occasions when we had come in contact with them. All six-footers and built in proportion, they were jet black with large, squat noses, flashing eyes, and fine white teeth. It was their laughter and cheeriness that got us. They had apparently been recruited as a labour battalion. Their officers were Tommies, their N.C.Os their own men. They spoke very little English; we believed them to be Africans from Bechuanaland, but weren't sure. They were well wrapped up in woollen scarves and Balaclavas and were dressed in the Tommy regulation battle-dress. They must have felt the cold terribly; it was bitter weather to us, and these men were obviously from an even hotter climate than ours. We admired the way they took it with unfailing cheerfulness. The boys told us there had been a panic among them when the first snows fell. Never having seen snowflakes before, they thought it was a gas attack and rushed to don their gas masks. None laughed so heartily afterwards as themselves.

On this particular day a score of them were digging a tank-trap some three miles north of our gun positions, just where we were to turn off to climb the mountains. Horrie spotted them and charged.

"Oh heavens!" groaned Don, "we're in for it now!"

There came a sudden howl, then a yell of laughter. We saw big black forms leaping down into the pit with a tiny white fury snapping at their heels as we ran up shouting at Horrie. But Horrie was determined to capture the lot; yelping hysterically he took not the slightest notice of us. His strategy was to drive every man down into the pit. When we arrived the pit was a roar of laughter with some blacks running up the sides of the pit while the men opposite ran down. The frantic Horrie raced round and round to chase back those men escaping. Other huge blacks leapt up on their friends' backs to escape the maddened Horrie who now would concentrate on one until he had him leaping back down the slope, then rush at another who was climbing up the opposite bank. Seeing these cheery chaps were simply revelling in the joke we sat down to regain breath and look on. They played up to him, never quite escaping, always on the point of doing so until the little fellow's legs began to wobble, his tongue hanging out, his eyes bleary, his bark growing less and less. All those big black heads, those great big eyes, those flashing teeth laughing up at him, drove Horrie to a frenzy of despair. When he got them all down into the trap he pantingly sat down but a huge black had crawled along the bottom of the pit and was crawling up at the other end, shielded by his mates' bodies. And Horrie with a yelp raced along the edge of the pit to block him. Then one came crawling up our side of the pit and as Horrie raced back to block him another was quietly sneaking up the opposite side. At last Horrie was hopelessly exhausted; when I picked the panting little wildcat up his heart was thumping with a speed that frightened me. The blacks came laughing up the sides of the pit and coming across to us a venturesome fellow reached out a huge paw to stroke Horrie. But a snarl and snapping of teeth sent his hand back at the double with a roar of laughter from his mates. The laughter made Horrie furiously angry, but there was nothing he could do about it; he was all in.

20

THE WOG-DOG ATTACKS THE WOLF

WE commenced to climb the mountains, Horrie insisting on leading the way.

"This climb will take some of the fire out of the little wretch," smiled Don.

"It would be funny if we really did meet a bad, bold wolf," I replied.

"I bet Horrie would have a go at him," declared Don.

Those mountains were deceiving. From the bottom they looked like a sloping wall, but now we saw there were numerous crests and small valleys between us and the top. A growl from Horrie attracted us towards two Syrian shepherd boys intently watching us. They took a great fancy to Horrie; his tiny size and uniform seemed to fascinate them. Despite their coaxing he refused to make friends.

Portion of these mountains were unscalable but to our surprise we saw in each little valley a number of villages. Some were so small as to be only a cluster of two or three huts of mud and rock, with flat roofs of mud and straw. On each roof was a small stone roller probably used to keep the roof flat and compressed, or to push off the snow after a heavy fall. In numerous cases the hillside had been excavated to form a side for the dwelling. The Lebanese villagers greeted us very hospitably; we were objects of considerable interest, particularly the ungracious Horrie. The villagers were proud Christian Arabs and would never ask for baksheesh. The men wore the traditional head-dress of sheepskin, coat, and long, wide pantaloons caught in tight below the knees. The women were shy but friendly. Numbers of these folk beckoned us inside to proudly show us their only pictures, those of Christ and the Virgin Mary. Their entire existence depended upon their small flocks of goats, long-haired sheep and pocket-handkerchief fields of wheat. We secured a bright little boy guide from one village and very proudly he introduced us to the folk around, to the undisguised resentment of Horrie. He showed us how the wheat is pounded into flour in an earthenware dish by rolling it with round stones like small cannon balls. The flour is then baked into pancakes. These, with olives and sour cheese made from goat's milk comprise their main diet. Very few of them could speak English but all could speak French, a result of the French influence in Syria since the last war. They liked the French. Little rock fences were built beside the villages for protection of their flocks against wolves at night. Each village possessed savage guards in the shape of huge, gaunt dogs that looked a cross between Alsatian and wolf.

"Ferocious beasts!" said Don distrustfully as they came snarling towards us. "Hold that little devil tight!" The Wog-dog was yapping in angry challenge. "The tips of all their ears are missing," added Don.

"Frost-bite maybe," I suggested. "Thank goodness!" for the villagers were driving the beasts into the houses.

With our guide in the lead we were climbing a wild mountainside when a deep-throated growl brought us to a halt. Barring out approach and with fangs bared stood a monster that was the nearest approach to a cross between a wolf and a lion I could have imagined. It completely ignored us; it was fascinated by the tiny Horrie who stood before it like a rat facing a mastiff. My heart was in my throat as Don's rifle slowly rose, while the little wretch advanced full of fight, determined to protect us to the death. He sailed straight into the attack followed by Don, the Syrian boy guide and myself. The mass attack staggered the wolf-dog who bolted with Horrie yelping at his heels. Horrie chased him; our yells to call him back only made him chase the harder; the huge dog easily kept just ahead of him, glancing undecidedly back now and then. The chase was downhill and the two dogs gained terrific speed, flying over the boulders and bushes round which we had to dodge. There was no chance of shooting the big dog without imminent danger to Horrie. My heart was in - my mouth as on dodging a boulder I saw the wolf-dog had turned and was easily leaping round and round the snapping Horrie. Evidently the huge dog was still undecided whether to annihilate this queer insect or not. The ground was so steep we could do nothing but hurtle ourselves straight at them as Horrie got a grip on the big fellow's toe. The latter instantly seized Horrie and shook him like a rat. We fell right upon the dogs, crashing and rolling over upon them. The big dog bounded up and was off like a startled lion, followed by a shot from Don which kept him moving. We saw his huge body vanishing amongst the bushes.

I picked Horrie up and examined him. He was nearly -all out, but it was only the wind knocked out of him. I'd thrown myself right at the little wretch. Feebly, he still wanted to chase the wolf-dog.

"If he hasn't had enough excitement for one day," panted Don, "then I don't know what we're going to do with him."

Horrie certainly was just about knocked out but for the remainder of the day he continued to bristle and growl to remind us we were quite safe under his protection.

It was several hours later that we stood on a mountain-top and gazed at the magnificent Valley of Lebanon far below - a valley where throughout history armies had marched and fought, and where were now two victorious Australian armies.

Wearily we climbed down; it was dusk before we reached the home of our guide in a very clean village. Outside our guide's home, corn had been spread on square sheets of cloth to dry in the sun; a very small girl was now collecting the corn. Her job of protecting it against the scrawny fowls finished with the setting of the sun.

We were made very welcome at the guide's home. Immediately on entering the room we gravely removed our boots and were motioned to sit at the fire.

This house, like all the others, was one large room. Conversation was mostly carried on by smiles and signs, but there was no mistaking the friendly atmosphere. But the little Wog-dog kept close by me and scowled at the whole Syrian family, eight of them. The evening meal was served as the sun went down. The table was only one foot high and as big around as a large dinner plate, so that the meal was "two at a time". They smiled to Don and me to gather round the table and make a start. The last two were mother and daughter. Dinner was a brown bread pancake, olives, and queerly tasting cheese. Under the watchful and approving eyes of the whole family we did our best and signed that it was lovely. To our dismay we were immediately rewarded with another and larger helping.

When the meal was over, cushions and Arab eiderdowns were taken from a comer and spread across the floor. Solemnly the entire family part undressed. We spread our-selves upon the eiderdowns, and were all soon fast asleep, except the distrustful, watchful Wog-dog. I doubt if he closed an eye all night. Every now and again I'd half awake to hear him growling by my side at any movement of the Syrian family.

Next afternoon we climbed down the mountain to be greeted by the waiting Imshi. Barking their felicitations, the two little dogs ran together. Then Horrie sat back and threw out his chest as they engaged in animated conversation.

"Leave them at it," smiled Don. "He's only telling her how he chased the wolf away."

We strolled across to the signal office where Poppa and Fitz and Gordie were deep in conversation.

"Hullo," exclaimed Poppa, "it's high time you two turned up. Have you heard the news?"

"No; what news?"

"Japan has bombed Pearl Harbour!"

"Phew!"

We stared; this was breathless news indeed, and our thoughts flew to our beloved Australia.

Japan and the Pacific and the U.S.A. were the sole topics of conversation throughout the camp, in fact through-out all Syria wherever troops were quartered. Almost every-one felt certain that the U.S.A. would defeat the Japs. In every camp men were speculating as to whether we would be recalled for the defence of our homeland.

It drew towards Christmas. Horrie and Imshi had a great time scampering through the snow. Horrie was much too "underslung" for the snow country; his long body on tiny legs sagged in the middle and nearly dragged on the ground. His best method of travelling now was by leaps and bounds; but often, only two bright eyes and a little black nose were all that we could see of the Wog-dog. The nights were bitterly cold and Horrie had an open invitation from all the Rebels as a footwarmer. To our disgust he turned us down flat, preferring Imshi instead. He had completely pirated Imshi by now. His warm lined box under my bed was scarcely big enough for himself but somehow or other he managed to squeeze Imshi into it too. The two funny little things, snuggled up under a rug, would sleep warm as two 'possums while we shivered. And despite our scoldings they would simply be there in the morning, gazing up at us with not a movement from either.

"Scandalous goings on!" growled the Gogg.

Imshi's master had given up in despair.

"She's made her own bed, she can sleep in it!" he declared. "I'd brought her up to be a nice, refined lady dog before that ruffian Wog-dog came along. Now that he's led her astray I wash my hands of her."

But the two little dogs did not seem to mind. Imshi daily imposed on her master's forgiving generosity by inviting Horrie to the anti-tank camp for their specially cooked meal.

Horrie would often return to the hut with his uniform sopping wet, then stand on his hind legs, and placing his front paws on a signaller's leg, vigorously shake himself, plainly saying "I'm cold and wet. Please take the coat off!" He'd rarely venture out into the cold without his coat. At night, it was hung up with Imshi's to dry on a rack by the heater. He wouldn't get up when we did but would stay under the blanket until we were about to leave for duty. Knowing we would not return for some time he'd warn Imshi and leap out of his cosy bunk. Then he would try to reach his coat, asking someone to put it on. His little growl of pleasure and tail-wagging and hand-licking were our rewards. When the thaw set in it was very hard on his pads and they became so sore that he could scarcely walk. Yet he refused to wear the boots we made him. He looked so mournful at being left behind in the hut that we made a sledge on which we could pull our signal gear and in the front of the sledge we built a little box for Horrie. Daily we had to walk some miles along the valley, checking the telephone wires. In his box under our warm greatcoats Horrie's little head would peep out. He loved this method of travelling, but coax as he would, Imshi would not travel with him but would trot along beside the sledge until we were just outside the camp area. Then she would trot back, Horrie watching her with that resigned "So long! see you later" expression.

On these cold walks Horrie loved us to pull up and boil the billy; he would have been a wonderful mate for a bushman. His "tea" though was a warm mug of Horlick's Milk which he quaffed with gusto. At times we had to leave the sledge but the gear was perfectly safe; no prowling native dare approach that sledge with Horrie on guard.

Horrie was a dinkum soldier-dog, always ready for duty. He had done guard duty for us in Egypt, Palestine, Greece, Crete and Syria, while on Crete he had done valuable service by carrying important messages swiftly down that big hill in the dark. On quite a number of other occasions he had carried out definite military duties. At this particular camp, we trained him within one hour to carry wireless news from the signal office to Mac's hut. The messages were placed in a leather tobacco-pouch used only for this purpose. When ordered "Take this message to Mac!" Horrie would take the pouch in his mouth and swiftly trot to the distant hut, quickly returning with the answer, if any. When he was on duty, even Imshi could not tempt him from his job.

We received the serious news that the Prince of Wales and Repulse were sunk.

Gordie rigged an issue wireless and with this we picked up Reuter's news in morse from London and passed it in sheets to the camp. And didn't the Wog-dog just come into his own as the bearer of the news to the eagerly wasting crowds.

We knew we would receive word to move any day. We hardly dared think of Horrie. The little Wog-dog had now been with us for so long, stuck to us through thick and thin, shared our joys and sorrows in sympathy and comrade-ship. In the sudden, and often unforeseen circumstances of war we knew a time might come when we might be compelled to part with him.

And he knew it too, knew as he always did when something deeply serious was afoot, and when something immediately concerned him. All were particularly tender to the little pup; often now he would gaze up into my face with such a searching expression that I instinctively knew he perfectly realized his fate was entirely in our hands.

We received the fateful news to move. And - what a relief - it was with our own friendly crowd. No change of commands, no order was given, no comment made about the little Wog-dog. Soon we were travelling back through Palestine. We felt sure it meant back to Australia.

We had received the bewildering news of the fall of Singapore, and of the loss of the 8th Division.

Horrie's farewell to Imshi was pathetic. We moved out before the A/T Regiment. Mac held Imshi up in his arms and Horrie gazed back until we were long out of sight.

They took us right back into Palestine and told us we would camp here for the time being. The "time being" slowly dragged on into eight weeks. Horrie fretted over Imshi for a time but eventually decided to buck up and take nourishment. The troops were very unsettled by the bad war news from the Far East, coupled with the inactivity and uncertainty of our own movements. Even leave now appealed to very few; all we wanted was to be on the way home. To pass the long days of waiting, the game of Racing Beetles spread like wildfire throughout the camps. Each man ambitious to race his own "Stable" hunted for the beetles around the countryside; there were countless ugly big black Scarab and other varieties. Quite a number of men grew so enthusiastic that they spent many patient hours training their beetles. To compete in any one race the beetles had to be the same breed, size and weight. Before each race, each beetle had to be first examined and approved or by the judge, "just in case" for we wouldn't put it past some of the lads "touching" up their fancies with a drop of liquid energizer, or other mysterious aids to speed. Any beetle that had been in the wars, such as a veteran with a toe nipped off or an eye missing was promptly declared a non-starter by the judge. If the suspicious judge detected a tiny speck of clay or anything else overweight attached to the undercarriage of a contestant, that beetle was immediately disqualified and the owner warned. The cost of entry for each beetle was two hundred mils. The owner of the winning beetle scooped the pool. The prize money for the winner of a big race such as a Haifa Doncaster or Palestine Flying or Jaffa Cup was really surprising.

As each beetle was declared a starter by the judge it was touched with a distinguishing mark such as chalk, red ink, grey paint etc., so that the punters could back their fancy and keep an eye on it during the race. And the punters eagerly followed the progress of their fancy; in a big field there were as many colours as jockeys wear in the Melbourne Cup. Some beetles were famous winners of many races such as the big brute Phar Lap, the nuggety Peter Pan, the swift Carbine, The Barb, Rogilla, Beau Vite, Ortelli, and numerous other lords of the Turf. The course was the centre of the ring of onlookers. The course proper was a clearly marked circle about four feet in diameter on the ground. Dead centre was the Starting Post. Here, the rim of a round cake tin was carefully placed, lid and bottom of the tin having been removed. Inside their barrier was placed the Field. Here they milled around all anxious to start or rather to escape while final bets were made. When all was set, the starter lifted the rim and away lumbered the beetles to all points of the circle hurried by yells of "Come on, Phar Lap!" "Come on Skipton, you beauty!" etc., etc., with groans when a fancied beetle collided with another and was knocked back several lengths. The first beetle to cross any part of the circle was declared the winner.

From the judge's decision there was no appeal, even though a dead-heat sometimes nearly caused a fight.

21

THE WOG-DOG IN DANGER

"WHAT on earth are you doing now?" growled Sergeant Poppa one day.

"Introducing you to Cuthbert Mark I," grinned Fitz.

Cuthbert Mark I was a queer chameleon lizard about six inches long. The weird crawlers looked like miniature prehistoric reptiles. Quaintly, one eye would sometimes look ahead while the other gazed astern. Each foot appeared to be a claw more suitable for tree-climbing than waddling across the desert. They moved in a slow, hesitating manner, placing one foot forward then lurching backwards and forwards before dragging up the other leg.

"Can't make up his mind," grinned Fitz as I held Horrie while Cuthbert Mark I did his stuff across the tent floor. He had a slick tongue as long as his body; he flicked it out and rolled it back like a youngster's "squeaker". He was an expert shot at flies, whether on the wing or as sitting shots. The tongue would flick out and with the same movement recoil back into the large mouth. Cuthbert Mark I would then squat solemn and motionless and lick his chops. There appeared to be a hard sticky substance on the end of the tongue and when this touched a fly it was hopelessly bogged.

"Just as well Murchie isn't here," sighed Poppa. "What with his craze for scorpions and lizards and beetles and asps, the army is turning into a menagerie."

The Rebels took to chameleon hunting. In short time Gordie produced Cuthbert Mark II, Feathers rivalled him with Cuthbert Mark III, and the tent seemed full of chameleons making life not worth living for the flies. The Rebel competitors earnestly stalked around the tent each with his chameleon perched upon the index finger, the chameleon as intent upon this satisfying game as the man.

On sighting a fly the man's finger would cautiously poke out towards it, the chameleon's tongue would shoot out and slide back with the fly. Angle shots, upside down shots, fluke shots, all manner of difficult shots were tried out to impress the spectators with the merits of Cuthbert Mark I, II, and III. Cuthbert Mark II proved the best eater; he could gobble twenty flies before his bingy was filled. Mark I could only manage twelve while Mark III lazily declined another victim after he'd gobbled eight.

When not doing their stuff, the chameleons were kept in a box at the foot of the tent pole. Fitz made Cuthbert Mark I earn his tucker by forcing him change colour often throughout the day. Fitz put such vivid blue and red colours into the box that Feathers protested, reckoned poor Cuthbert would kill himself turning himself inside out trying to keep pace with the changing colours.

Horrie grew very jealous of these strange pets; often I would catch him gazing longingly at the box but I would shake my head reprovingly. Poor little Horrie exerted his will power and I could almost see him sighing, but he did not touch the box. Until --

Early one morning I noticed that Horrie had a somewhat hangdog look about him. As I glanced at the chameleons' box, Horrie rolled over on his back with a guilty invitation to play. Fitz sat up yawning, noticed my face and the box.

"Finish Cuthbert Mark I?" he inquired.

"Finish!" I replied.

"Finish Cuthbert Mark II?" came Gordie's voice.

"Finish!" I replied.

"Finish Cuthbert Mark III?" inquired Feathers.

"Finish!" I replied.

"You cannibal!" scolded Feathers, but Horrie rolled on his back and wagged his tail.

Time went on. It seemed certain we would return to Australia. Then word was whispered around that lest animal or bird diseases be introduced to Australia all pets must be destroyed or left behind when we again moved camp.

"What are you going to do with Horrie?" asked the Wog-dog's numerous friends.

"I don't know yet," I could only reply.

Something must be planned quickly.

Horrie gradually lost his enthusiasm for parades, even for play. For days on end we were compelled to force him leave the tent even for exercise, all he wanted to do was lie on the end of my bed with his big brown eyes staring voicelessly. That dog knew.

The boys were constantly dropping into the tent for a chat. It got on our nerves the way each man would unconsciously lower his voice, pat Horrie, and say "Poor little Horrie," or "Poor little Wog-dog."

Horrie would watch us talking, his head on his front paws, his eyes sad and searching.

One day Feathers remarked, "I'm getting that way that I can hardly look Horrie in the face."

"Same here," said the Gogg.

As the Gogg spoke, Poppa tramped into the tent.

"This joint is like a morgue," he growled. He picked Horrie up and continued, "even the little Wog-dog has the miseries." He gently stroked the Wog-dog. The Rebels remained silent.

"It is a shame," growled Poppa, "to leave the little fellow behind. I well remember just before we returned home from the last war the Light Horsemen and En Zed soldiers shot their horses rather than leave them in the hands of the Arabs."

"Better than abandoning them to living death under the Arab whip," scowled Fitz. Horrie nosed my arm.

"What's the matter, little dog?" I asked and knew that could Horrie speak he would answer "Please don't leave me behind."

"What would be our position if we were caught trying to smuggle Horrie home?" inquired Don.

"If we could actually get him home, maybe we could get him quarantined, then he'd be all right," answered Poppa, "but if discovered before arrival he would be destroyed."

"If we leave him behind it means the Arabs or the desert," said Fitz. "Give him a fighting chance."

"He is entitled to his chance," declared Gordie. "He's, stuck to us through thick and thin."

"Give him his chance!" said Don.

The Gogg walked up beside Poppa and patted Horrie. "Cheer up, little dog! We won't let you down."

"Well, that's decided," said the relieved Gogg. 'We'll have to be careful now! One little slip and ---!"

"There'll be no slip," I said, "but it means very careful thought and teamwork from now on."

The first thing we did was to sneak Horrie away to Tel Aviv for veterinary examination, just to satisfy ourselves that he had no disease which he might introduce into Australia. Sergeant Poppa wangled the necessary leave for Don and me. We found a vet., a kindly, grey-haired man speaking fluent English. A refugee Jew forced to fly from Germany, he and his family had to start life all over again in a strange land. In his neat little room packed with medical books and photos of animals he examined the strangely quiet Horrie, whose eyes never left my face. The vet obviously knew his job, and was gentle too.

"He appears quite healthy he declared, "but I would like you to leave him with me a week until I can make tests and be sure."

To Don's doubting glance he smiled. "He'll be quite all right. Come with me and meet some of my little patients."

We followed him to the back yard which was divided into small compartments completely covered with a wire fence. Here were the patients, one little fellow with his leg in splints. One glance showed us the dogs were well cared for and had no fear of the old doctor.

"Your little dog can remain here," he said and pointed to a vacant run.

I lifted the uneasy Horrie into the run. "Good little dog," reassured Don and patted him. I could hardly look at the little fellow's eyes.

"Can you leave some personal article until you return for the dog?" asked the vet.

Thinking he meant a deposit and not holding too well I started to remove my wrist watch.

"No," he smiled, "a sock would serve the purpose admirably."

"A sock?" I inquired.

"Yes. He will treasure an article with your scent upon it and will thus expect you to return."

I removed my boot and handed the sock into the cage to Horrie.

"He will be reassured now," said the vet. "Say your temporary farewell."

We did so, trying to comfort the silent and miserable little Wog-dog.

"Anyway we're doing the job conscientiously - and he's in good hands," I said as we walked back along the street to pick up the leave bus.

"Yes," answered Don. "By the way, it was a good idea about the sock."

"Yes, but don't tell the Rebels or I'll never hear the end of it."

We trudged on silently and I wondered if Horrie felt as comfortable with the sock as I felt uncomfortable without it.

That week was a drag to all the Rebels, we realized how very much we now would miss the Wog-dog. On the third day the Gogg said "Go and ring up that vet, and find out how the Wog-dog is!"

For the life of us neither Don nor I could remember the vet's address.

"Nit-wits!" exclaimed Poppa fiercely.

At last the day arrived. Poppa wangled it so that I must drive a truck into Tel Aviv for "urgent military stores". And that truck was in a hurry. The old vet smiled when he opened the door.

"The little dog is quite all right," he said. "Come in."

With a sigh of relief I found myself again in the little room. "Come and we'll get him," invited the vet.

I called Horrie as he opened the back door. There was a flurry like sudden wind amongst leaves, and Horrie was leaping up frantically, nearly wagging himself in half. The vet. retrieved the sock from the kennel.

"Do you require this?" he asked.

"Guess I'll leave it," I laughed. "Did it do it's job?"

"Most certainly. It was a great comforter. That sock has remained with the little dog all the time. If he walked in the yard for exercise he carried the sock with him. It was in his mouth when he returned to the kennel. He slept with his head upon it."

Back in the room he said, "I have been totally unable to trace any signs of canine disease. Your little dog is in excellent health. Should any military necessity prevent you taking him with you I will readily accept him."

"All Hitler's armies could not prevent us taking him away," I answered happily. "At least there will always be one of us left to look after him."

He smiled and patted Horrie.

"So I should imagine," he said gently, "and I know a people who would love an army such as yours. You have a very faithful little friend here," and he patted Horrie again.

Despite my insistence he gently but firmly refused a fee. At the front door I asked, "What breed is Horrie?"

He smiled and simply answered, "Just a nice little dog."

"Well, we don't care, do we, Horrie?" I asked the Wog-dog.

Plain to see the Wog-dog now did not have a care in the world.

Sincerely wishing the kindly vet the best of luck the little "no-breed" dog and I walked out into the sunlight and the truck. Horrie put his head out the window and joyously barked at everything we passed.

Back at the camp Horrie's excited bark brought Don to the tent door with the Rebels behind him. No need to tell them the great news.

"You beauty" was the chorus.

"I was positive he was healthy," declared Feathers. "Now we know for certain."

"No. 1 hurdle over safely," laughed Gordie. The Wog-dog was trying to jump all over him.

"Good show," said Poppa approvingly, "and now I've got some news. Big Khassa the L.A.D. dog was taken for a ride this morning."

Ominous news.

"So you'd better not let Horrie out of your sight," added Poppa Sand - you'd better make up your mind quick and lively just what you are going to do with him."

We thought out plan after plan. Once the dreaded order to destroy pets came out we knew it would be rigidly enforced. Days went by and we discarded plan after plan. It was by no means simple. As the authorities knew how not only we but the whole battalion were attached to the Wog-dog, it would need no ordinary plan to smuggle him home. The day we learned the A.A. Regiment had received word to move out we assembled in the tent and carefully fastened it against intruders.

"Our turn to move next!" said Poppa grimly. "We've got no workable plan thought out. Now let's get to it - or we lose the Wog-dog."

For hour after hour we debated it. One suggestion was to say he was run over by a truck. But if they dug up his grave there would be no body. Another was to keep him hidden in an abandoned trench a mile from the camp until we moved out. But someone might stumble across the trench. If we said we had given him to some English people in Tel Aviv then the authorities could check the address. As we found faults in all our schemes, we realized with a queer tinge of distress how difficult it was going to be. And this meant only the concealing of Horrie until we left this camp. There would be other camps then and the voyage to Australia.

"The most urgent thing is for him to disappear before the actual move," I said desperately, "for us to get him to a safe place away out of camp so that no inspection can possibly locate him."

"It will have to be some place where we can pick him up at a moment's notice," said Gordie, "for we may get two hours' notice to move in the middle of the night."

It was Fitz who gave us the first glimmering of a water-tight idea.

"What about saying that we gave him to a Palestine policeman in Tel Aviv," he suggested, "but we don't know his address. Horrie could then be in any one of the numerous police block-houses in Palestine. The authorities would hardly be likely to check them all."

"Not a bad idea," said Poppa thoughtfully, "but the authorities will be very suspicious of that excuse."

"I have it!" I exclaimed, "we will actually give Horrie to the Palestine police I will get one of our own military police to accompany Horrie and me to the police block-house on the beach near Ascalon. The provost man will be an irrefutable witness."

"But how are you going to get him back?" demanded the Gogg.

"I'll sneak him back!" I declared.

"How?" demanded Poppa.

"I'll give the dog to the police. Then return here and report to the authorities that we've given the dog to the Ascalon police. A few hours later, I'll hurry back to the police and tell them I want to borrow the dog for a few hours to take a last photograph of him with his unit. And then - I'll forget to return him."

"And if we are ordered to leave at a moment's notice?" demanded Poppa.

"Look here," I said, "could you manage to detail me for duty with a truck early tomorrow morning?"

"If it was for urgent military duty," replied Sergeant Poppa.

"Very well, it is for very urgent duty. I'll take Horrie and the witness with me and hurry into Tel Aviv. I'll find the policeman all right; those English police are fine fellows. I'll bet the very first man I ask will willingly offer the little dog a home. Then I'll hurry back, drop the witness at headquarters and that very night will have Horrie back here in camp again."

"Which leaves only one problem," said Poppa dreamily. "Until we move we must keep Horrie perfectly hidden from everyone in camp."

"The tent is the safest," declared the Gogg, "but we can't keep him here. Although the authorities would know the Palestine police have taken him over, still there will be tent inspection in routine and Horrie would be discovered."

"It's a pity we could not safely hide him in the tent somewhere though," said Fitz thoughtfully. "If we could establish his alibi by giving him to the police, then keep him hidden amongst us in this very tent the whole problem would be solved."

"Look!" exclaimed Don and pointed to the cane mat that now served me as a bed. "A hole under Jim's bed!"

"You beauty" came the chorus.

If we could hide Horrie underground in a hole under the mat on the floor then the closest of tent inspections would not find him.

22

HORRIE IN PERIL

"To work, my hearties!" cried Poppa as he dragged away the mat. "We'll dig the hideout all ready for tomorrow night."

Rows of tents were in front of ours, and to each side. But our tent was one of the back row and there was only desert behind us. Poppa kept watch outside the door, Fitz squatted down with a book outside the tent, Gordie sat idly whistling on the side opposite. I sprawled out at the rear of the tent. Inside, Don, Feathers and the Gogg got to work. Now and again a low "all clear!" would come from the three watchers. I would reach in under the tent bottom and Gordie would place a bucket of sand to my hand. Cautiously I would hurry to empty it in an old disused trench at the rear, then hurry the bucket back. We must not leave a pile of damp sand outside the tent.

It meant a considerable number of buckets and there were numerous interruptions for of course men were moving about all the time. In a few hours the boys had dug a hole five feet long, five feet deep and two wide in the tent floor where my mat should be. We lined the hole with packing case boards and at last in pleasurable triumph gazed down at a deep, secure comfortable hideout. When the sleeping mat was pulled over the hole there would not be a man in ten thousand who could have the slightest suspicion of what was underneath.

"All we've got to do now is to train Horrie to stay quiet as a mouse in his foxhole," chuckled the Gogg.

That night after lights out, with Horrie coiled asleep as usual at my feet, I lay awake planning details of the coming morning's venture. There must be no mistake.

Immediately after breakfast I set out for the police block-house near Ascalon, but without the military policeman. The block-houses were built at vantage points across the desert, to be used as strong posts in the event of a native uprising. Large, square concrete buildings two or three storeys high, they were believed capable of withstanding a siege of twelve months High above each was an observation tower giving a wide view over the countryside and also ensuring communication by signal with the next block-house in the event of telephone lines being destroyed. Surrounding the approach to my blockhouse was a barbed-wire fence. A policeman was busy washing down a police car; he glanced up as I approached.

"What is it you want, Aussie?" he inquired.

"Well," I grinned bashfully, "I was wondering if you would do me a favour.

"Certainly, if I'm able to."

"Thanks. It's this way. In an hour or so I've got to drive our major to Gaza. After dropping him there, well, the truck's mine for the rest of the day so long as I don't drive straight back to camp, of course. My mates and I want a last fling in Tel Aviv. But here's the snag. The major always takes his little dog in the truck with him and I'll be expected to take truck and dog straight back to camp. I don't want to take the dog to Tel Aviv lest he gets lost or someone steal him from the truck. So I was wondering if I could leave him here and pick him up on our way home tonight?"

"You Aussies always seem to be A.W.L.," he smiled.

"It's our last flutter in Palestine," I grinned hopefully. "All is set, but I wouldn't like to lose the major's little dog. He's a good sort."

"What time would you be returning?" he asked.

"About 9 o'clock tonight."

"I will be working here throughout the day only. However, when I'm finished this evening I'll ask the duty man at the gate to hand the dog back to you. Will that suit?"

"It certainly would, and thanks very much."

"Very well. When are you bringing the dog?"

"About ten-thirty."

"Very well, I'll be here." And I drove happily back to camp.

"It's all set!" I told the waiting Poppa. "Now contact the provost sergeant straight away. If you can persuade him to send a provost with me in a truck to hand Horrie over to the Palestine police, then all's well."

"I'll fix it," declared Poppa.

"Right. And spread the news that Horrie is being handed over to the Palestine police as a mascot."

"You bet I will."

I hurried to the Rebels to report progress. The Rebels scattered to spread news of Horrie's fate, while I took Horrie on a lead to Captain Hindmarsh and Big Jim to tell them the little dog was on his way to a new home. They were very sorry to see him go, but agreed with me that it was far better than destroying him.

"At least he'll have a good home," said the captain sympathetically.

It was only when they were farewelling Horrie that I imagined I detected a faint glimmer of suspicion in the officers' expressions, so I hurried to the waiting truck lest I say or look too much. The Rebels and quite a few other lads had already gathered at the truck to see the little dog off, others were coming from among the tents, they were strangely silent. I caught a puzzled questioning look in face after face. I thought we'd better push off before the crowd gathered. We lumbered away to shouts of "Farewell

Horrie", "Good-bye Wog-dog", "Good luck, little fellow". Then came an oppressive silence with only Horrie barking.

Ron Ford, the truck driver, was very quiet; he was an old cobber of mine but I dare not tell him or anyone about the scheme; we had made an unbreakable rule that for perfect safety no one must know but the Rebels. Of course, the provost in the truck was not in the know either.

The truck pulled up outside the police fence and I jumped straight down with Horrie and hurried inside lest the policeman still washing the cars should stroll out to the truck. The policeman ceased work, nodded and said, "Tie him up here," indicating a horse trough. I tied Horrie to the trough.

"He will be quite all right here," promised the policeman.

"Thanks very much. I'll call back for him tonight."

"Good-oh. Have a good time and don't get into trouble!"

"No fear of that," I grinned, "it takes me all my time getting out of it. So long for the time being."

"Cheerio."

I hurried back to the truck with Horrie's entreating barking ringing in my ears. I turned and called to the policeman.

"Take good care of him, please!"

He waved his arm in reply.

"Let's go," I said and climbed into the truck.

As we drove back to camp Ron said, "Bad luck, losing Horrie after all this time!" The provost also sympathized with me. I mumbled something about "He'll have a good home, anyway," and lapsed into moody silence.

On arrival at camp I found Horrie the sole topic of conversation; the unbelievable news had spread that Horrie was handed over to and accepted by the Palestine police as a mascot. Captain Hindmarsh and Big Jim unconsciously helped our plans, declaring it an accomplished fact, as did Ron Ford and the provost. They had seen the Wog-dog handed over and accepted.

"Just in the nick of time!" Poppa hoarsely whispered to me. "Orders are out! All pets must be destroyed!"

This news helped us face the boys, for some of them did not take the news of the Wog-dog's departure too well at all. Cal Taylor, the pay-sergeant, demanded, "Don't you think you at least could have tried to get Horrie home?"

No excuse would satisfy Cal, plainly showing by his manner he believed we had let Horrie down. Major Haupt sympathized with me at the loss of the little Wog-dog, while Big Jim told us it was the topic of conversation during the evening meal at the officers' mess. Bill Cody, one of the signallers, came up to me and demanded, "What's this I hear about Horrie being given to the Palestine Police?"

"Yes, that's right, Bill."

"Then it's a darned shame; I'd have tried to get him home if you'd given him to me!"

"No go, Bill; he would surely have been destroyed if caught and you know how strict the order is. As it is, he's alive and in kindly hands."

"Are you sure he will be well looked after?"

"Certain. They're Englishmen, you know."

"I suppose he'll be all right," he said doubtfully, "but I would have given anything to see that little dog get home."

Scores and scores of the troops came to us to learn the truth about Horrie; some were outspoken in their opinion that we should at least have tried to get the little fellow home. We could see that many were not happy about it at all though they grudgingly admitted it was far better that he be in kindly hands than destroyed. We sensed that others secretly thought we had let the little dog down.

The desire to tell them the truth was almost over-whelming, but one whisper might spoil all. We took all that was coming to us and were not happy about it, keeping up the deceit for Horrie's sake.

Now, the secret is out at last. Very many soldiers who still think Horrie is a mascot of the Palestinian police will be happily surprised to learn he is at home in Australia.

But the getting of him there! That evening I slipped away from camp, a towel over my shoulder as if going to the shower down the road. At 9 sharp I was outside the police fence. I whistled, and a moment later a little white shape came racing to me fast as his tiny legs could carry him. The policeman had unleashed him the moment I whistled. I picked up the Wog-dog and laughingly trying to quieten his embraces hurried back to camp. Quite a few shadows were coming and going among the tents so I wrapped Horrie in the towel and whispered to him, "Lie quiet, lie quiet!" Swiftly but cautiously approaching the tent I was glad to see the rubbish can was in its usual place; had it not been there it would have meant we had visitors. Inside, the Rebels were anxiously waiting and I almost spoiled everything by slipping Horrie loose. But Don's hand snatched over his mouth just as he was about to bark.

"Sssh!"

"Sssh!"

"Sssh!"

Horrie cocked his head to one side looking very concerned, trying to understand what it was all about. We worked hard on him; obediently he settled down, big brown eyes asking us what was next. We were about to try him in his hideout when the Gogg who was keeping watch at the door hissed "Gow". Instantly I scrambled under a blanket making a tent with my knees, with Horrie wedged between them. Big Jim popped in with cheery greeting and sat on Don s bunk.

"And what's the matter with you?" he said down at me, "on your sleeping mat in bed at this time of the evening?"

"I've got a touch of sandy fever," I sighed.

"Heavens! For goodness sake, don't get sick now!"

"I'm quite all right," I said hastily. "Will be O.K. by morning.

"You'd better," he said grimly. "It would be awful if we had to leave you behind."

"I'll be jake!" I declared and quietly spread my hand down over Horrie's nose.

"I wonder how poor old Horrie is tonight," mused Big Jim. My hand had tightly closed around Horrie's nose, my knees around his body in the nick of time to stay the violent wriggle at the well-known voice and mention of his name.

"Oh, he'll be all right," I said and wearily closed my eyes. I dare not glance at the faces of the Rebels.

Don sauntered to the door of the tent drawling, "I guess he's pretty miserable, poor little Wog-dog," and sauntered on out of the tent.

What's the matter with Don?" inquired Big Jim.

"Upset over Horrie," answered Poppa.

"It s a darned shame!" declared Big Jim. "Horrie was a great dog, a great -"

"Ssh!" I said and sneezed to the convulsion between my knees. "Let's not talk about it," I added weakly and turned over.

"Well, I think I'll be turning in," declared Sergeant Poppa and led the way to the tent door. The Rebels solemnly agreed they might as well turn in also and Big Jim rose yawning to his feet. He wished us all good night, and followed Poppa out the door.

Breathlessly we listened until the crunch of sand died away, and then doubled up in suppressed laughter.

"For heaven's sake, don't laugh!" came Don's voice as he reappeared at the tent door. "This thing is serious!"

Presently Poppa reappeared, having seen Big Jim to his tent.

"I can't stand too much of this," he grinned, "but what is more I object strongly to being a party to deceiving a very good superior officer!"

"I wonder really what Big Jim would think if he knew the truth about Horrie?" said Feathers. "Couldn't we let him in on it?"

"No!" replied Poppa. "He'd be as right as rain, but it would not be fair to him. Orders are orders whichever way you may look at it!"

We were all in perfect agreement. "I bet he'll laugh when he finds out, all the same," declared Don. "And now we'd better train Horrie to his hideout."

The Gogg kept guard again outside the tent door-while we rolled the matting back and introduced Horrie to his new home, all nice and cosy with what old clothes we could spare. We lowered him down into it saying "Good dog, good dog!" but Horrie thought it was a new game and tried to jump up while urgent hands reached down to smother his bark. We "Ssshd!" at him until in sheer disgust he lay down on the old clothes. "Good dog, good dog," we assured him and rolled the matting back over the hole When I'd spread my blanket over the matting it was pretty dark down there. There wasn't a whimper from Horrie; he went to sleep.

At reveille, we let Horrie scamper around inside the tent for exercise.

"You'd better suffer a relapse of that sand-fly fever and stay in bed all day," ordered Sergeant Poppa as the boys filed out to the parade ground. "Meanwhile, continue with the training of Corporal Horrie for if you think you are going to malinger here doing nothing you're jolly well mistaken!" and the old warrior tramped out to parade.

So back Horrie went to his dugout while I "Ssshd" him, and soothed him all through the morning. He was a bit restless down there, trying to understand why he should not be out on parade on such a beautiful morning. I removed the tin drinking-dish as he bumped it now and then; such a noise could give the secret away. By and by he quietened under the "Sssh!" at regular intervals.

The supreme test came later when I heard approaching footsteps.

"Tent inspection!" I thought. "Heavens!"

And in marched Captain Hindmarsh with Ron Flett, the sergeant-major.

"Hullo, Moody," said the captain, "what is the matter?"

"A touch of sand-fly fever, sir," I replied.

"Better take it easy for a while," he advised.

"Very good, sir."

The inspection was quickly over. The Rebels purposely had left the tent unusually tidy with all gear stacked in correct place. The orderly officer, with a nod to me marched out followed by the N.C.Os.

When they'd stepped outside the tent I heard a distinct dog's yawn down in the depths below. Horrie had remained quiet as a mouse.

23

HORRIE TAKES IT ON THE CHIN

THE Rebels brought our food from the cookhouse, but Horrie was not allowed his bones until lights out lest visitors hear the crunching.

"Well now," said Sergeant Poppa that night, "how are we going to transport Horrie when orders come to move off?"

We decided on a pack-bag. I lifted Horrie from his hideout, covered him with a towel, then Don and I sneaked away with him out the back of the tent. We hurried away to the vacated camp area where the ack-ack regiment had been and put Horrie to the ground. He hardly knew what to do with himself and scampered round and round us in delight. To please him we played chasings but when he barked in excitement we sat down until he'd worked off his energy. When he came trotting back to us we gave him his bones; then his training commenced in earnest. I put him in the pack-bag on Don's shoulders but for a start left the flap open. Then away we went, tramping round and round the old camp area. I walked close behind Don, talking reassuringly to the Wog-dog whose wise little head gazed at me from the open pack.

Now and again we'd stop, lift Horrie to the ground and pat him and whisper "Good little dog, good little Horrie, very smart pup!" And he just lapped up this praise. I changed places with Don and away we went again. Every time Horrie moved Don whispered "Sssh! still, still." Apart from the effort of learning what all this meant Horrie was not comfortable for his weight in the bottom of the pack tended to make it fold in on him. We did the best we could.

After several hours of this practice we fastened the flap. Horrie was then completely encased in the pack-bag. Again we marched around the area. Horrie now remained very quiet but the tip of his nose appeared where the flap just covered the side of the pack. Gently pushing back his nose, I whispered "Quiet! good dog! good dog! sssh!"

For a start, we left him in the closed pack only five minutes; then we'd undo the flap and let him poke his nose out. As a final lesson on that first night, we put the covered pack on the ground while Don and I sat beside it and every time he moved whispered "Sssh! still, good dog!

Thus, night after night we trained Horrie until he would stay two hours in the covered pack, quiet as a sleeping mouse. And now when we showed him the pack, he would quietly step into it and cuddle down on his own accord. Day by day, for eight long days he remained noiselessly down in the deep black hole. As to tent inspection now, Big Jim asked with a puzzled frown, "What on earth has come over you Rebels? Have you reformed?"

"Yes," we grinned.

Big Jim could not understand the uncanny tidiness, the scrupulous accuracy of gear. One glance now and inspection was over so far as the Rebels' tent was concerned.

Then came the great order. We were to move out in two days' time. Tomorrow there was to be a trial parade, each man with all gear moving out of camp precisely as he would do for the last time on the following day.

This meant a problem to the Rebels. We dare not leave Horrie in the hole next day, for every man must move from camp; every tent would be stripped bare. Under such conditions, tent inspection must discover even a button upon the bare floor.

"Horrie must attend parade with us," declared Fitz, " In his pack!"

"He'll stand up to it all right," said Don confidently.

And he did. But we had not allowed for daylight and the heat. Horrie had been trained in the dark and cool of night. He now needed all his stout little heart.

We moved off with Horrie in the pack on my back, the Rebels keeping an eye on the pack from the following ranks. We'd missed out very badly in foresight, for every man's pack was bulging with kit, while mine looked as if it had collapsed, was only half-filled. Luck was with us; none of the heads noticed. Horrie, very hot, only moved once then became instantly still at Don's warning "Sssh!" After the parade, when back in the tent I let Horrie out as quickly as possible; he was nearly smothered. We put him back in the hole and I got down and whispered to the panting pup. He licked my hand to show he trusted us implicitly As I laid him down on the old heap of clothes and stroked his head he sighed and dozed off to sleep.

We set to work to improve the pack. We cut round holes in it to admit air, camouflaged the holes with netted string, lined the pack with plywood to keep it square and give it the solid appearance of a tightly filled pack. Then we fitted the pack on Don's back. It now looked the part of a well-filled pack and, what was just as important, it was much more roomy and comfortable and admitted much more air. We rolled a pair of socks very tightly and fastened them behind the pack. This kept the pack out nearly an inch from Don's back which did away with considerable heat and also allowed air to flow between Don's back and the pack. We then put Horrie in and were delighted at the improvement; he could really move now and the slight movement would not be noticeable. We fastened the flap, then fastened the roll of blankets on it, with the tin hat held between the two straps in the centre of the pack. We stood back then and saw it would be impossible for anyone to guess there was anything but a man's kit inside the pack. We stepped out into the night and gave it a thorough tryout. It was perfect.

When back in the tent, we took Horrie out of the pack and his pleased expression and tail wagging emphatically assured us "It is O.K. now!"

As at dawn we must pull down the tents and pack them away, we must quickly find a new hiding place for Horrie on this last night. We solved the problem with the aid of the canteen sergeant, a good sport. He hid Horrie away in a cupboard of the canteen for the last day and never breathed a word about it.

That night I recovered Horrie and quietly edged away from the roadway, sleeping well away from the troops, and joining them in the morning with Horrie in the pack as they were crowding into the transport buses en route for Gaza. When there, Horrie had to lie quietly for many hours in that pack before the train arrived. The Rebels then shepherded us towards the guard's van, and before the guard knew what was doing, Don and I had hopped in and slammed the door. We were determined about it but presently managed to square the annoyed guard though he nearly threw a fit when we let Horrie out of the pack and Horrie immediately tackled the worthy Oriental gentleman. It proved quite a comfortable ride for Horrie during all the long trip through the Sinai desert back into Egypt.

With Horrie again in the pack we filed out of the train at Kantara on the Suez Canal and marched to a transit camp. The pack was quickly put to the ground and surrounded by the Rebels' gear in such a way that Horrie still had air through the string laced holes. It was fearfully hot. As we ate, I noticed Captain Hindmarsh now and then glance in our direction and laugh. I felt uneasy but am sure he did not suspect anything, he was always a cheerful O.C. When we moved off the Rebels managed to stick together. We marched to the Canal, with Horrie silent as a mouse, to the heavy "tramp, tramp, tramp" of feet. As I marched on I knew how the little Wog-dog would dearly love to be marching at the head of the column. But it simply could not be. At the Canal we boarded punts, crossed the Canal, and once again entrained.

We had to travel in a crowded carriage, bound for Port Tewfik. It must have been misery to Horrie, the tramp of feet, the marching songs, the laughing voices throughout the train journeys, all the sounds of an army on the move, that was sweet music to him and now throughout it all he had to huddle in silence in the depths of a black, suffocatingly hot pack. Don fought for a window seat and we were thus able to place Horrie's pack with the airhole in the back of the pack facing the open window. Thus at last he got a cool breeze. He needed it. If only the Rebels could have taken possession of the carriage, we could have let Horrie out of the pack. But we daren't risk it; these noisy boys were not in the know.

Night came. The boys stretched out in the best way they could. Silence at last except for an uneasy snore, the squeak and rattle of the carriage wheels, the dreamy hum of the train. I placed the pack on its flat side so that Horrie could lie down and poked my fingers through the strings that covered the air hole and stroked his head. He gave a doggy sigh and then remained quite content. In the small hours we arrived at Port Tewfik, and in the bustle of detraining Don carried the pack. We formed up and marched to a transit camp where the boys simply rolled up on the sand and slept the last few hours to dawn. Just before dawn I sneaked right away from camp with Horrie and finding an old trench let him loose there, waiting until Don should get the hang of the new camp. Horrie quietly slipped away to do his little business and then came racing back up and down the trench to stretch his little legs. I sat there for hours, puzzling and wondering, discarding plan after plan as to how I could get him on the ship.

Don found me about 9 o'clock. He had managed to locate the Rebels who seized their opportunity to secure a tent to themselves. He then located me and I carried Horrie under a blanket through the camp to the tent. We took it in turns to watch at the door while again and again I had to grab Horrie and slip under the blanket as lads came in to gossip. The lads were so restless they could not remain still.

At midday Sergeant Poppa came into the tent. "We march to the embarkation point tomorrow morning," he announced.

"Hooray!" we shouted. "So we really embark."

"Sure thing. But I've heard another whisper!"

"Out with it, you grizzled old oyster."

"Imshi!"

"What?"

"Yes. The Anti-Tank Regiment has just camped not far away. One of the boys whispered to me that so far they've managed to smuggle Imshi all the way from Syria to Port Tewfik in a truck."

We were delighted at this news of Horrie's girl friend; his little tail would have wagged itself right off could he have seen her again. Regretfully we decided we dare not risk it. We were almost on the ship now and it would be heart-breaking to lose the little dog at the last moment. Love must wait. Apart from unwise risks, the greatest problem of all lay just ahead of us - how to get Horrie aboard ship. The Rebels sat around in serious conclave.

"If there's one slip," said Poppa, "it means the end."

"It will be the most severe inspection of all," declared the Gogg.

"I've thought out a plan," I said. "Here it is. The troops will be formed up on the wharf in three ranks. After 'Open order, march!' kit-bags and packs and gear are placed at the owners' feet. Now, when the inspecting officers draw near us, I'll quietly faint. Feathers and Don will break ranks and come to my assistance. But on reaching me I'll weakly regain my feet with their assistance. Sergeant Poppa noticing the disturbance will hurry along and advise Corporal Feathers, 'Help him up the gangway and get him to a cool place!' Fitz will then immediately move to my gear and say to Sergeant Poppa, "I'll take his gear aboard, Sergeant." "Yes," Poppa will reply, "do that." Fitz will then pick up the pack containing Horrie and escort me up the gangway. Once aboard Feathers and Don will inquire where I'm to bunk, meanwhile I will have recovered sufficiently so that there'll be no need to take me to the hospital bay. We'll just slip down below and find a bunk or a latrine where I'll wait with Horrie until all the Rebels come aboard and sort things out. Now how's that for an idea!"

"Excelsior!" they declared admiringly.

"You might even think your way out of jail," said Poppa grudgingly.

"He'll probably have to one of these days," declared Fitz.

Under cover of darkness that night Don and Feathers and I took Horrie out on the desert. While he exercised his legs we sat a long time under the brilliant stars. This was to be our last night in the Middle East.

"I wonder what the future has in store for us," mused Feathers.

"Goodness knows," said Don.

We lapsed into silence, wondering what the future meant for Australia. We were very unhappy. Singapore gone, Java gone -we did not dare to think too much.

Next morning the camp was in a bustle. Soon, we were on the march, Horrie in the pack on Don's shoulders, and I marching behind to cheer Horrie with a whispered word. Feathers and Fitz marched at Don's left, Gordie and the Gogg on my right.

That day, the sun blazed down upon a desert that radiated suffocating heat. Horrie suffered terribly. He made no movement, uttered no whimper. At half-way, we halted for a few minutes. The Rebels screened me while I wet my fingers from the water bottle and thrust them into the top of the pack to feel Horrie's famishing little tongue.

"Good dog, Horrie," I whispered, "good little pup! Stick it out!"

The last mile dragged by in silence but for the muffed tramp, tramp, tramp. Sweat poured down faces and necks and bared chests brown with dust. I leaned forward and put my ear to the pack. I could hear Horrie panting, but he never moved. He was a little Anzac; the breathless desert sun poured down upon that suffocating pack but could not break his spirit. He would have died without a whimper. At last, at long last, we marched out on to a wharf; soon now would end one of the greatest triumphs of endurance that ever a brave dog went through. On the wharf we sat -wearily down to wait. I carefully lifted the pack from Don's back and placed it with the hole in the back facing a cool breeze from the water. The Rebels squatted around while I again thrust wet fingers into Horrie's hot prison. I whispered "Good dog!" and could hear his little stub tail brushing along the bottom of the pack. Towards anchored ships heavily laden ferries were taking troops. In low voices we were rehearsing the fainting stunt when Sergeant Poppa located us, his grin as wide as the funnel of a ship.

"There's to be no kit inspection on the wharf," he chuckled in a low voice. "Not until we board the ship, anyway."

This was a relief as great as it was unexpected; it meant I would not have to become an amateur Clark Gable. A ferry came bustling alongside, we crowded aboard, carrying Horrie in his pack. There was only standing room. I gained the side of the ferry and stood with my back to the water. Thus the pack was out of sight of the crowd and also received the breeze. All hands were eagerly staring across the harbour wondering which ship was to be our particular transport. Pleased were we when our ferry headed towards a mammoth liner; excited too, when we saw the Stars and Stripes lazily fluttering over the stern, our first glimpse of the United States Navy. She was the first United States ship to carry Australian soldiers in this war. She was now named West Point, and was formerly the S.S. America built in answer to the Queen Mary. Cheery greetings floated down to us from the Yankee sailors. The Rebels crowded round me as we climbed the gangway and into the ship through a big iron doorway in the side. Horrie's lucky star was in the ascendant; there was no word of the dreaded inspection. We were immediately given a ticket and found ourselves in a huge cabin that could accommodate all the Rebels and our cobbers Reg Jenks and Ron Baker as well; it was really a palatial flat. Of all the luck! We shut the door, lifted off our packs and grinned delightedly at one another. In the bustle and excitement and anxiety this really seemed too good to be true. Our extraordinarily large, luxurious cabin also had a separate shower room all to itself, this room was also fitted up to the Nth degree as a princess's boudoir - or Horrie's suite.

"Just watch this closely!" I whispered mysteriously to Reg and Ron as I carefully undid Horrie's pack.

"First roll up your sleeves," said Don earnestly, "just to show them there's nothing there."

I did so, made a few flourishes over the pack and said "Hey Presto! Sir Knight arise from the Wars!" and out popped Horrie.

"Well, I'll be blessed!" exclaimed the dumbfounded pair.

"I don't know about 'blessed'," smiled Don, "but we'll be damned if we're caught."

Horrie was all wags and grins and about to bark that he was no phantom dog when a command from me quietened him on the instant.

"But I thought he was with the Palestine police!" exclaimed Reg.

"How on earth did you manage it?" demanded Ron.

"It's a long story," I replied. "We must make Horrie secure first. In the meantime, not a whisper about him."

"You can count on us," they promised in one voice.

Horrie was in a bad way; to feel him was almost like touching a hot-water bottle. Don and I gave him a refreshing bath and he brightened up immensely after Feathers mixed him a cool drink of Horlick's Milk. The Rebels allotted me the bunk with an air inlet above it and I placed Horrie on the mattress under the inlet.

As the cool fresh air flowed over him he seemed to droop wearily, gave a few sighs and in a moment was fast asleep.

"The poor little Wog-dog is all in," sympathized Gordie.

While we stripped for a bath I quickly told Reg and Ron the story. We were so delighted with our success so far we would have loved to tell all the boys on the ship. But this could not be, for it would mean the end of Horrie. The secret must never come out until he was safe in Australia. We wondered if the Anti-Tank Regiment was aboard our ship and, if so, whether the men had managed to smuggle Imshi aboard. We would soon find out; it would be wonderful if the doggie romance were to have a happy ending.

24

THE STOWAWAY

"BETTER work out the plan of campaign now," advised Gordie, "before we're called out on duty or parade or goodness knows what."

"We must keep Horrie hidden below," said Fitz. "The lock on the cabin door eliminates a surprise entry; we must always keep that door locked."

"One man should always remain in the cabin," I suggested, "and we need a code knock so that the man inside will know whether it is a Rebel or stranger who demands entry.

"Suggest a knock," said the Gogg.

"How about W.D. in Morse?"

"Why W.D.?"

"Because it stands for Wog-dog."

"W.D. in Morse is the knock," they agreed.

"So that," suggested Gordie, "if a stranger knocks at the door the man inside pops Horrie into the bathroom before opening the door".

"But what if the visit proves to be a surprise cabin inspection," said Don, "and the inspecting officer wishes to see whether the shower room is clean and tidy?"

"Tell him that someone is using the closet," replied Gordie.

"What if he insists - despite the bloke inside?" asked Feathers.

"That's the danger," I agreed. "The word privacy has no connection with a private in the Army."

"Well, what do you suggest?" asked Fitz.

"My only suggestion is the pack," I answered.

"That might do," said Gordie, "but you must keep this pack always handy with the gear always neatly stacked on the bunk. Then you could whisk Horrie into the pack, "Sssh!" him as you buckle the straps, and place the pack in its place with the gear on the bunk. Then yawn as if you'd been asleep when you open the door."

Thus was decided the plan of campaign.

Don and I remained in the cabin while the boys strolled out to have a look round the ship and fish for news. I patted Horrie gently but he was dead to the world.

"He's all in," said Don. "I would never have thought that a dog could willingly have put up with all that he has.

"It seems a long time ago now," I replied, "since we left Syria."

"Yes. But with a bit of luck I think most of his troubles are over."

"If we can keep him hidden during the voyage then there'll be only the last hurdle to overcome, smuggling him ashore in Australia."

The W.D. knock came softly at the door. We unlocked it and in stepped Sergeant Poppa. His first glance was toward the Wog-dog.

"That little pup has more guts for his size than I thought possible," he said, "but his troubles are mostly over now. We're not wasting any time; the ship pulls out tonight. No lingering about for days, and no convoy. One thing about the Yanks, they don't mess about."

"Where are we bound for?" I asked.

"Don't officially know, but am pretty sure it's Australia." Just then the Rebels knocked and were admitted. Their faces were wreathed in smiles. "We're sure going to little old Australia, buddy," Fitz informed us.

Everyone felt quite sure, though it was still only guess work. But next day when well at sea we joined an excited crowd around the ship's notice board.

"It is expected that this ship will arrive at Port Fremantle at 10 a.m., 26 March 1942."

The news flew through the vessel and a dense crowd came pouring down to the notice board. There never could have been a happier ship-load of soldiers.

"Glance at their faces," grinned Poppa. "Ask any one of them to lend you ten bob and he'll give you his deferred pay.

It was later still when the Rebels were gleefully discussing the news in the sanctuary of the cabin that Sergeant Poppa came in.

"I see Corporal Horrie is fit for duty again," he grinned and patted the Wog-dog, who now was his old cheery, tail-wagging self. "Well, boys, here's what's doing. There will be no cabin inspection today. Here are routine orders from tomorrow onwards. The cabins will be inspected about 10 a.m. During inspection one man to each cabin remains below in charge. The remainder must be up on deck and are not allowed to return below again until 12 o'clock."

"I'll stay below each morning so as to be near Horrie," I volunteered. And this was agreed to.

"The captain of this ship is the most popular man in the Red Sea," resumed Poppa.

"Why is that?" inquired Don.

"His nickname explains it," answered Poppa, "they call him 'No Parade Kelly'! There will be no parades at all aboard this ship."

"You beauty came the chorus. "Three cheers for No Parade Kelly! He'll do us!"

"I'm going to like the Yanks," declared the Gogg. Just then a strange knock came at the door and I snatched up Horrie while the Rebels broke into laughter and song. To a "Sssh!" Horrie snuggled quiet as a mouse in his pack bag. Poppa opened the door. Big Jim stepped in. He was so pleased he didn't notice the delay in opening the door.

"Well, I suppose you're happy now!" he grinned. "How is your cabin? Comfortable? But I don't suppose you'd mind if you were parked below in the coal now you're going back to Australia. Great news, isn't it?"

"Old stuff!" I answered.

"How's that?" he demanded.

"We knew it way back in Syria."

"Oh yeh!" he grinned. "Wise guys, eh?"

"No; just plain privates," Fitz answered solemnly.

"Well, what about a little celebration when we land back in Aussie?" suggested Big Jim.

"Right. Let's make it a surprise party," I replied. The Rebels instantly guessed what the "surprise" would be. Big Jim glanced in a puzzled way at their laughing faces.

"What's the surprise?" he asked. But they only grinned at one another and winked mysteriously.

"How about doing some tricks for us at the party?" Reg asked me.

"Right," I replied, "but it will be one trick only, a magic trick especially for Big Jim. On one condition!"

"And what's that?" asked Big Jim.

"That afterwards you will tell no one the trick."

"But what is this mysterious trick?"

"Can't tell you now, but it is called 'Pack up Your Troubles'."

I thought Sergeant Poppa was going to burst.

"Well, I don't know what it is all about," said Big Jim, "but I certainly feel interested."

"You'll feel bewitched," declared Poppa, "when you see the trick."

"I'll look forward to it," smiled Big Jim.

When Big Jim returned to the deck we let Horrie out of the pack.

"How would you like to be a little surprise dog, Horrie?" chuckled Poppa. "Now you are here, now you vanish! You've had ample practice, anyway."

It was quite O.K. with Horrie.

Next morning brought inspection. The Rebels had the cabin shining like a new pin. Then they trooped up on deck. I felt quite confident. This would be child's play to Horrie.

"Stand by for cabin inspection!"

With a final "Sssh!" to Horrie I stood to attention as into the cabin trooped the O.C. of our unit, the company commander, one of the ship's officers, and a few N.C.Os thrown in to make up weight.

They gave the cabin a pretty good once over, too. They even peered under the bunks. To me, they seemed peeved they could find nothing to complain about, so they had to troop out. I let them get well clear before I took Horrie out of the pack.

"Very good dog, Horrie!" I told him. But his expression was lackadaisical; cabin inspection was puppy stuff to him.

The next day passed cheerily until a pressingly serious problem upset us. Horrie had been well trained; in the desert or in Greece or anywhere at all he would never dream of making his private arrangements anywhere near our tent or camp. But here he was unable to leave the cabin and he was fighting against nature; it was the second day before we noticed it. I took him to the shower room and tried to encourage him; I stayed with him many hours, trying to suggest the little sunken pit below the shower. Unknowingly, I was defeating my purpose by remaining with him.

"Any luck?" anxiously inquired the Rebels who now were very concerned.

"Nothing doing," I replied gloomily.

Poor Horrie was now drooping, obviously unhappy, and very unwell. Fortunately I woke up to the fact of his shy-ness so put him in the shower room, stepped out, and closed the door after me. I opened it half an hour later and the dejected pup crawled out looking guilty and miserable.

"You poor little blighted I sympathized, "but you are a good dog really."

But he was still very unhappy. It was not until I left the shower running and closed the door after me that he realized he was not going to be scolded. He gazed up with brightening eyes, his tail commenced to wag, and all was well.

But, several days later came a shock to our happily organized cabin. Sergeant Poppa brought the news.

"What's the matter?" I asked at his serious face.

"A lot," he answered gloomily. "We must act quickly. A mistake has been made. It appears this cabin really belongs to another unit; you are ordered to vacate. Worse till, there is absolutely no hope of getting such another cabin to yourselves now."

"'Struth! Wouldn't it?" exclaimed Gordie.

"Where are we to be placed, then?" I asked.

"There are bunks empty in Headquarters Area," he replied, "but they are only in ones and twos," and he produced a plan showing the cabin numbers and empty berths.

We spread out and scouted round in very uneasy mood lest the fact that Horrie was on the ship might now get around. They were all good coves, we knew, but suddenly to produce the little Wog-dog after all this time - well, one unguarded word, one slip of the tongue would mean the end of Horrie.

With a quickening of hope we found that one large cabin contained members of the Signal Platoon - Bill Arrowsmith, Bill Cody (one of Horrie's especially close cobbers), Syd Jordan, "Dar" Davis, Gordon Baxter, Bill Martin, Bob Groll. These men had all known Horrie well. I was certain that no thoughtlessness on their part would betray him. There was one berth vacant in their cabin. Returning to the Rebels, I inquired what they'd found.

They'd located empty berths here and there, but all agreed it would be best for me to doss in with the signallers.

"We'll pop into your cabin daily and take it in turns to watch over Horrie while you take a breather on deck," said the Gogg.

So we set about the unpopular job of moving our gear. It was stiff luck indeed that we who had stuck together so long should be separated on the very last phase of the campaign, the voyage home.

With Horrie in the pack-bag and the Rebels giving me a hand with my gear, I entered the new cabin. The signallers were there. I locked the door, then solemnly I undid the pack and out popped Horrie on the floor.

In amazed silence, they stared at the phantom. It was not until Horrie ran with wagging tail to his old friends that a delighted yell arose.

"Sssh!" I exclaimed with upraised hand.

They still could hardly believe their eyes. With laughing faces they crowded round Horrie who in his own doggie way was busy showing how well he remembered each and every one of them. Finally they demanded the story. For a dog to appear like this when they one and all believed he was far away with the Palestine police must be a story worth hearing. It was an appreciative audience with the little Wog-dog sitting back listening to every word, now and then signalling approval by a cute glance at one of the boys, as if to say "Now, what do you think of that!"

They agreed that Horrie was as game as Ned Kelly.

They all agreed that we should carry on as before - keep the door locked, use the W.D. signal, I to remain below each morning for cabin inspection. But they wouldn't hear of the Rebels coming along to relieve me for a stroll on deck. Horrie was an honoured occupant of their cabin now and they themselves would take it in turns to relieve me. The air inlet in this cabin was over Jacky Gardner's bunk.

"Horrie can sleep on my bunk under the inlet," said Jacky. And I was glad, for we were approaching the Line and the heat in the cabins was already almost unbearable. Horrie had "fallen on his feet" again.

But he had one bad week ahead of him. It was breathlessly hot in the closed cabin. He never complained, but would life his nose to the inlet to get any breath of the hot air. We dare not let him out of the pack until after cabin inspection. Then I would sit him on the tiled floor under the shower and fan him with a towel until he cooled down. But when I stopped fanning, his coat would become hot again. Every half hour I'd wet him and his tail-wag was my reward. After midday, when the boys came filing down to the cabin, there were plenty of willing hands to carry on with the fanning. The West Point was a palatial ship but had been designed for the Atlantic crossing and was not at her best in tropical areas. For six days, until we'd crossed the Line, Horrie suffered considerably but soon afterwards became his old tail-wagging self.

Up on deck the days were glorious; the Aussies and the Yanks got on famously. The Yanks soon took to the great national pastime, two-up. They were great sports - cheerful winners, good losers. Games, sports and competitions helped further to cement an already strong friendship. Even the ship's padre brought up a laugh. In his sermon, he said that Moses had left his disciples in charge of the food supply, to wit, a quantity of fish, very scarce at the time. Solemnly warning the disciples not to touch the fish until he returned Moses went about some particular job. But the appetising odour of the fish aggravated the disciples. One by one, those fish disappeared. "When Moses returned," concluded the padre, "boy, was he mad?"

The Anti-Tank Regiment was not on board our ship.

We used to wonder whether the men had got Imshi aboard ship and whether they would manage to smuggle her into Australia. We determined to try and get in touch with them should they land at Fremantle at the same time as ourselves. It would be the end to a perfect romance if we only could bring Horrie and Imshi together again.

The Day came at last, the Great Day. Excited shouts, coo-ees, then cheer after cheer rang through the ship. We raced to the deck, and there, slowly but surely out of the haze, our own dear land was becoming visible. There could never be anything else so good!

And then tragedy. And it was tragedy.

Word flew round that "Hobo" was aboard and - they were after Hobo.

Hobo was a beautiful cat, C Company's cat, a mascot and a lovely animal. He had been born during the evacuation from Dunkirk, with death raining from sea, air and land, with thousands of men dying, in a sea of sinking ships and drowning men.

Later, the Tommies had proudly presented Hobo to C Company.

They became greatly attached to that lovely, proud and s intelligent cat. He had accepted their company and had accompanied them through everything - everywhere. He went with them through each campaign by land and sea, in peace and war. And they had smuggled him aboard. We did not even know.

And now, through some tragic oversight, some tiny piece of carelessness, right in sight of Australia, it had become known that Hobo the cat was aboard.

The order was that he must be destroyed.

C Company refused to give up the cat.

Hobo could not be found.

The ship slowed down and stopped. Silently, thousands of men gazed towards the dim shores of Australia.

The ship would not proceed until the cat was produced.

Hobo's friends held out. We wondered if we could hold out so long if it had been Horrie that was discovered. Anxiously we waited. Thousands of men were waiting.

Not one man even whispered that the C Company lads should hand over their cat.

They held out for twelve long hours.

Hobo was produced. Thrown overboard.

The ship gathered way, gained speed.

At last we pulled into Fremantle, the wharf was a sea of cheering faces. A car with a charcoal burner caught our eye, the first we had seen.

"What is it?" someone yelled, "a travelling bath-heater?"

The good-humoured owner put a match in somewhere and out shot a flame to roars of laughter from the troops.

"Let's sail back to the Middle East - it's too risky here!" A wag yelled.

How clean and fresh the houses on the hill looked and, yes, there was an old-fashioned square, two-storeyed pub with handkerchiefs fluttering from the balcony.

"I'll bet it's called Bay View," laughed someone.

Yes, it was Australia all right.

"If the Japs ever land in Australia," said Poppa, "they'd better bring a multitude - they'll need them!"

A dog was trotting up the road.

My heart beat fast. Horrie would soon be doing the same with just a little more luck.

25

THE AUSSIE-DOG

THE West Australian boys disembarked, and happy lads they were. Next morning the ship was under way again, bound for Adelaide. Just a few more days and Horrie for the first time would put his little feet upon Australian soil; his eyes after many days would again see the open daylight.

We were very anxious. We never left Horrie an instant and kept that door well and truly locked. The tragedy of Hobo weighed heavy upon our minds.

During those few hours in Fremantle by circuitous means we had tried in vain to hear news of Imshi. Evidently her ship had not yet called in.

On that last night aboard, the Rebels with Sergeant Poppa crowded into the signallers' cabin to make plans for bringing our long laid schemes to a happy ending. Horrie was so excited he nearly wagged his tail off.

"Horrie is quite confident," laughed Feathers.

"He's a cert!" grinned Gordie. "He'll make history in Australia yet."

"A dinkum emigrant," smiled Fitz.

"He'll pull his weight," growled Poppa. "What's country without a dog anyway."

We all agreed.

"Now then, you nit-wits," said Poppa, "there's to be kit-inspection on the ship tomorrow morning. So you can hardly make a mess of it at the last moment. However, there may possibly be an inspection on the wharf. I doubt it; the Army is too excited at setting foot on Australian shores again. Now, just in case of an inspection on the wharf, what is the plan?"

"The fainting plan," I said.

The Rebels and sigs were confident this would succeed although I would probably be chucked into an ambulance. This would not matter. I felt certain I would be able to hop out with Horrie. The Gogg gave me the address of his people in Adelaide. "If anything happens and we become separated," he said, "take him straight there. He will be safe as a bank with my people until you can claim him. If you like, you can leave him there for good."

Good old Gogg. But it was my secret hope to leave him with the old folk in Melbourne.

"Well now," said Feathers seriously, "you know how we all feel about Horrie, Jim! But he is more your dog than anyone else's; at least, the little Wog-dog has adopted you a shade more than anyone else. We can't all have him. He's yours. There's no need to ask do you want him for keeps."

"I do," I said, "but I'll make you all a promise. First of all, I promise he'll have the best home in the world. Secondly, I feel confident that some day he may take upon himself a bride and I only hope she will be Imshi. In either event, I promise you each a puppy."

Loud cheers. "I'll call my little bloke 'Horrie's son'," declared Don. Each of the Rebels quickly thought of a name. But when the signallers claimed a pup each I felt the least bit doubtful. I knew Horrie had the heart of a lion but whether he could manage a harem - well!

Next morning we anchored at Port Adelaide. Big Jim was the officer at the last cabin inspection. Never had cabin been so spick and span, nor so quickly inspected.

Horrie was as quiet as a mouse in his pack.

"Good show," said Big Jim approvingly, and disappeared to inspect the next cabin. Then lining up in single file, Rebels and signallers before and behind me we shuffled-our way along the passageways until we gained the gang, way. Horrie was in the pack upon my back. Never did smuggler's heart glow as mine, even though he carried a: fortune in diamonds upon his person. All I carried was a little dog.

When I reached the gangway I laughed in noiseless mirth. I wanted to shout for sheer joy. There was no inspection - the troops were filing across the wharf and on to a waiting train.

Almost in a dream I scrambled into a carriage. Rebels and signallers had grabbed a window seat for me. Very; soon, from within the pack bag Horrie was gazing at Australia through a crisscross of string.

Happily we enjoyed the tea and cakes given by a Women's Welfare League. Then we were moving towards Adelaide. Few knew of our coming in those terribly anxious days and we were wildly cheered by all who saw us.

At Adelaide, we were marched to temporary billets at the Adelaide Oval. The Rebels came across to Don and me and quietly shepherded us away behind a grandstand to allow Horrie a few precious minutes' freedom. Horrie stepped out of the pack, stood quite still a few moments gazing up at the wonderful sky, his feet upon Aussie land, his little nose sniffing the sweetest air upon earth. Then with a delighted yelp he raced away to investigate his first dinkum Australian gum-tree.

Horrie was acclimatised.

Horrie was whisked back into the pack as special trains came to take us to Burnside Camp.

A pleasant Australian woman holding a bunch of papers approached Don and me.

"Billets for two," she smiled. "Come along."

With a "Good luck!" from the Rebels we followed her. She took us to a pretty little house. We waited inside the gate until she knocked at the door.

"What if they don't like dogs!" whispered Don.

'We must chance it," I whispered, "but I think everyone likes a dog."

"Would you come in, please," called our escort. "This is Private Gill and this is Private Moody," she said to a decidedly pretty young woman. "Mrs Trezona."

We smiled our best and were shown into a clean little room.

"I hope you will be quite at home here," she said anxiously.

"It's great," said Don, "it will do me."

"Me too," I murmured.

"Unpack your things and then join me in a cup of tea," she smiled.

As the door closed behind her Don said quietly, "She seems a real good sort."

"Yes," I answered, but we won't produce Horrie until we're sure."

Presently, she called us to the kitchen.

"I hope you don't mind having tea with me here," she said.

We assuredly did not mind. It was a very pleasant little kitchen. We got along so famously that at last I said with an anxious grin, "Are you fond of dogs?"

"I adore them."

"Excuse us!" said Don and I as we jumped up together and foolishly collided in the doorway. I can still see the surprised look on her face. More so when we reappeared wreathed in grins carrying a weather-stained pack bag.

I quickly opened the flap and a little white, inquiring head poked out; two large brown eyes gazed hopefully up at our hostess.

"The dear little war-dog!" she exclaimed and we knew Horrie had captured yet another heart.

When Mr Trezona came home that evening we told them some of Horrie's adventures. The party that sat down to the evening meal was Ted and Mary, Don and Jim and Horrie.

The little Wog-dog was going to love Australia and Australians were going to love him.

Next day, when Don and I met the Rebels on the parade ground, our broad grins told them the news.

"He's right as the bank," I added. "Taken possession of the place. I managed to get a wire home to Dad in Melbourne too. Horrie and I arrived in Australia.'"

The next few days were joyful ones for the troops so long on foreign soil. The Adelaide folk did all they could for us, while we were so happy to be in Australia again that the hours just melted by. Our unit was to be entertained at the Burnside Town Hall and Mary suggested we invite the Rebels home afterwards for a surprise party.

This was just the very thing; in delight we arranged the surprise we had promised Big Jim.

After that very merry dance, the Rebels gathered at the Trezona home and filled the place with noise and laughter. Here we were, with the exception of good old Murchie, all united again in an Australian home.

The laughter and chatter were interrupted by Fitz standing up and calling.

"Attention everyone please! The prime object of our happy meeting tonight is to present to you all - and especially to Big Jim - the world's greatest magician!" and he bowed to me. I solemnly reciprocated. "Also," proceeded Fitz, "introducing to you the magician's famous assistant. Dot and Dash," and he waved towards Don and me. Solemnly we acknowledged the hand-claps.

"The trick you are about to witness," proceeded Fitz, "has never been performed before and never will be again. To achieve the phenomenon about to be witnessed Dot and Dash have blended the elusive magic of the mysterious East with the skill of the cultivated West, backed up by the initiative and determination of our own Sunny South. Time is passing, the Moving Finger writes on, the curtain rises upon the Past to bring to the Present the magic you now will see." And Fitz sat down amid thunderous clapping.

Don and I rose to again take the bow. I held up my hand.

"Ladies and gentlemen, particularly our guest Big Jim, in whose honour this miracle is about to be performed, I first ask you in all seriousness to swear you will not say one word of what you are about to witness - not until at least forty-eight hours from now. Do you all swear?"

They swore unanimously.

"Very well. Silence!"

Don vanished. In a now expectant silence he reappeared reverently carrying an object covered with a white sheet. Carefully placing it in the centre of the carpet he made a few passes over it then stood silently.

I bent over the sheet and made mystic passes murmuring "Antibrecazzar! Antibrecazzar!" then in a sepulchral voice said, "I now ask our friendly-enemy officer, Big Jim, to step forward and remove the covering. Then to open that which waits underneath."

Big Jim was laughingly thrust forward; somewhat doubtfully he gazed down at the sheet.

"I'll be the goat," he laughed. "I'll get even with you on parade afterwards."

Slowly he pulled off the sheet, a pack bag was underneath. With a puzzled smile he undid the flap, opened it -

For a second he stared, then - "Horrie!" he shouted. "Well, I'm blessed!" and Horrie was barking up at him in a frenzy of tail-wagging.

Pandemonium.

As Big Jim lifted up the excited pup, the girls crowded around to pet him - the pup, I mean. Big Jim would have appreciated it much more; Horrie passed me a scornful look. "Trying to make me a sissy," he seemed to say.

When Big Jim called for silence he simply said: "Friends, words cannot convey what I mean when I say this has been one of the very pleasant surprises of my life. you could only understand had you been over there and known how this game little Wog-dog entered into all our lives. To thus meet him again on dear Australian soil, well -"

Next day the Rebels got a whisper about Imshi. The rumour made us feel unhappy. She was discovered, so the rumour said, only one day before the boat reached Fremantle. She disappeared.

That same day I was en route for Melbourne. In the quick movement of troops to their home towns, and in the soon-to-be quick movements to the danger now hammering at our very own North, troops lost touch with one another, and to this day we are not really certain as to Imshi's fate. It was the hardest of luck that her friends should get her to within one day's sail of Australia and then be forced to destroy her. But we still believe she may be happily in Australia, her friends when she was discovered may have "worked a ready" and got her ashore after all. Anyway, if Imshi's owner reads these lines and if Imshi is in Australia he will now know there is a little Wog-dog called Horrie very anxious to meet his lady love again.

Let us bring them together again, for the sake of the old romance.

There is not much more to tell. I got Horrie safely home to Melbourne where he is very happy with the old Dad, a better man than I. We, the Rebels and the signallers and the battalion in our wanderings since have missed his cheery company very much. But we all look forward to the reunion, the great reunion when the Wog-dog and Murchie and we all meet again after the war.

Until then the little Wog-dog sends greetings and wishes for a safe return to his very many friends. The little out-cast of the

Desert, Corporal Horrie of Egypt, Greece, Crete, Palestine, Syria, is a dinkum little Aussie-dog.

April 1944. Much water has flowed under the bridge since our great party. Since those fearful days we and our great Allies have driven back the enemy from Australian shores and our dear country breathes again. There is much, alas, very much more to do yet, but how immeasurably stronger we are to do it than we were in those fateful days when the Wog-dog first set foot upon Australia.

Since then, we have been through the hell of New Guinea. But what I am in a hurry to tell you all is this. Only a few days ago, Don and I, back from New Guinea on leave in Melbourne, heard a shout across the street. He was one of the boys; he had just been landed after a long period of guerrilla fighting in the East Indies. He dare tell us but very little as you can readily understand. But he told us this great news. Murchie is still boxing on. Murchie is safe, fighting as we knew he would, to the last. He is away up in the mountains and chief of half a dozen villages. He has his own little commando, he is still waging his private war against the Jap. He will hold out, will Murchie. Soon we will meet Murchie again.

Hurrah!

EPITAPH

Well, Horrie, little fellow, your reward was death. You who deserved a nation's plaudits, sleep in peace. Among Australia's war heroes, we shall remember you.

Under Quarantine Regulations, Horrie was destroyed on 12 March 1945.

ETT IMPRINT has the following ION IDRIESS books in print in 2023:

Prospecting for Gold (1931)
Lasseter's Last Ride (1931)
Flynn of the Inland (1932)
The Desert Column (1932)
Men of the Jungle (1932)
Drums of Mer (1933)
Gold-Dust and Ashes (1933)
The Yellow Joss (1934)
Man Tracks (1935)
Over the Range (1937)
Forty Fathoms Deep (1937)
Madman's Island (1938)
Headhunters of the Coral Sea (1940)
Lightning Ridge (1940)
Nemarluk (1941)
Shoot to Kill (1942)
Sniping (1942)
Guerrilla Tactics (1942)
Trapping the Jap (1942)
Lurking Death (1942)
The Scout (1943)
Horrie the Wog Dog (1945)
The Opium Smugglers (1948)
The Wild White Man of Badu (1950)
Outlaws of the Leopolds (1952)
The Red Chief (1953)
The Silver City (1956)
Coral Sea Calling (1957)
Back O' Cairns (1958)
The Wild North (1960)
Tracks of Destiny (1961)
Gouger of the Bulletin (2013)
Ion Idriess: The Last Interview (2020)
Ion Idriess Letters (2023)

www.ingramcontent.com/pod-product-compliance
Lightning Source LLC
Chambersburg PA
CBHW030232170426
43201CB00006B/194